I0137389

Spotting Fakers, lies, and illusions using elementary theories about the mind

Discover how to give your lie detector a permanent turbo boost

by

Miklos Zoltan

Although the author made every effort to ensure that the information in this book was correct at press time, the author does not assume and hereby disclaim any liability to any party for any loss, damage, injury or disruption caused by errors, omissions, instructions, advices, recommendations, suggestions or any other contents of this book, whether such errors, omissions, instructions, advices, recommendations, suggestions or any other contents of this book result from negligence, accident, or any other cause.

Names, characters, businesses, places, events and incidents are either the products of the author's imagination or used in a fictitious manner. Any resemblance to actual persons, living or dead, or actual events is purely coincidental.

You (the reader) are fully and solely responsible for your own choices, actions, and results. Do not read this book if you disagree with any of the above.

4rd Edition

To my wife Catherine -

who helps me find the way.

Thank you.

Contents

COVER ART DESIGN

Malraux Zoltan

EDITS

Nancy M. Padrón

Why Read This Book

Subjects covered:

- Spotting Fakers
- False-information
- Fundamentals of the mind
- Practices

Fakers are individuals who covertly plot to take advantage of us. *Spotting Fakers* offers information for identifying them, which allows us to minimize the adverse effects they have on our lives. The proposed methods for dealing with them are simple, compassionate, and strictly nonviolent.

False-information is a general term, which refers to all forms of incorrect, incomplete, or missing information such as lies, incorrect assumptions, errors, mistakes, omitted information, etc. The increased ability to spot false-information in any field can have a positive effect on our ability to solve problems. False-information can affect all fields of knowledge from gardening to astrophysics.

Fundamentals of the mind fills in missing information we may have about the workings of the mind. Since all of us receives one of these (a mind) at birth, but without a user manual, it is common to have unanswered questions on the subject. Discussing the nature of knowledge (how we know), answers many of these. The presented theories focus on why we are susceptible to false-information, manipulation, and influence and how to free ourselves from these.

The *Practices* section offers simple exercises for spotting false-information and fakers. We all have the natural ability to spot false-information but certain factors can cause us occasionally to override our judgments. The practices can help restore our ability to spot negative influences and increase our trust in our split second judgments.

Here is why you should read on:

At first look, these subjects may appear a bit involved, and something only those who specialize in the related sciences should read. The use of simple language, and everyday examples we can relate to, make these subjects easy to digest. Persistent problems, assumptions, betrayal, and false-information affect most of us at home as well as at work. We can feel the adverse effects of these as a mental load or burden in our daily lives. Like a splinter in our mind. Freeing ourselves from these is fun and exhilarating.

The sources of such mental loads are hidden from us, which is the reason why we have them in the first place. Since they are hidden from our sight we don't know how many there are until we start looking for them. Doing that is relatively easy, and the rewards are well worth the effort to find out how. The results are increased freedom from negative influence, manipulation, and persistent problems.

Introduction to fakers

The word "faker" is essentially a label. People are individuals and not labels. This book is not a promotion of label identification or labeling. The word "faker" allows for simple writing about the subjects and phenomena related to faking. Even when we suspect someone to be faking, it just means that, we may want to examine very carefully whether to trust him or her, or not. The purpose of this book is to help with such decisions.

Labeling people as "fakers" aloud is *not* what this book promotes. Nor is it about thinking of people as just a label – faker. No matter what the actions of an individual are they are still individuals and this book promotes views with the intention to help see people even more clearly as individuals as opposed to just bunching them under a label.

When we see someone cheat or betray others, we tend to assume that the bad deed was due to a temporary lack of judgment. In some cases it is. In many cases however, the bad deed is the result of careful plotting and scheming. People who behave this way are fakers. As part of their scheme, they pretend to be trustworthy, helpful, and friendly. They can do this for a long time because it is part of their character. This is what they do. This is their true talent. They may *appear* to be a trustable professional but when the time is right, they betray those who trusted them.

We fall for their devious trickery for simple reasons. We don't know what is going on inside their minds. What we see, is what they show us by facial expressions and body language. They smile, act calm, and flatter us. Since we can't read their thoughts, we build our trust on what we see. The longer they keep at it the more likely it is that we will accept them as trustworthy. They can

3

pretend for years or even decades because for them, that is natural. What we cannot see is that behind their smiles and expressions there are inconsiderate and harmful intentions. For the rest of us, it is hard to imagine that someone could be that way. The majority of us are unassuming in this aspect. Even when we have initial suspicions about a person, we tend to ignore our doubts over time as the acting continues.

Fakers are extremely persistent and this gives them an advantage. The purpose of this book is to even the odds. Fakers represent a minority but can seriously undermine societies from the inside out. Since faking is what they do best, they tend to rely on that skill mainly. As a result, most of them have serious flaws in their claimed profession. Therefore, they tend to do more harm than good.

Fortunately, their mentality gives them plenty of character traits, which we can recognize. At any given time, there are millions of fakers waiting dormant for the big score while the rest of them are poisoning our society with daily unethical conduct. Some incidents make the news but most never receive due attention.

The nature of fakers is not a widely known subject and isn't considered essential knowledge for everyday living. This book is an attempt to change that consideration.

Many of us at some point in our lives suffer the adverse effects of fakers directly, while the rest of us only see the indirect results. Ailing economy, corporate and government corruption, and poor service affect all of us. To a great degree, fakers cause this decay. Widespread awareness about them can greatly help reduce these negative effects.

No known field or profession is free of them. Even science and education has just as many fakers as any other field. As a result, they spread their false-information and false views in these fields as well.

Spotting Fakers

Only if we had a built-in-lie-detector we wouldn't need to bother with the signs of fakers, right? After all, they are essentially liars. Not just any liars but compulsive ones. They also have hidden plans to exploit and betray. This is important because those who lie but never harm others are of no consequence. Those who lie while deceitfully seeking advantage are of great concern.

Interestingly enough, we *do* have a built-in lie detector. However, we tend to override our instinctive thoughts when fakers continue to feed us more reassuring lies, using sweet talk and flattery until we give in. When we learn to spot fakers what we are essentially doing is, restoring trust in our split-second judgments. Fakers are essentially illusionists. They can be excellent at creating the false impression of trustworthiness. This is not surprising when we realize that most of them practice their faking skills for life. Of *course* they are good at it. Spotting fakers is relatively simple and the following chapters describe the telltale signs. The tricky part is to accept and trust our judgment. Later chapters reveal why we tend to override our instincts and fall for the illusions. This can help recover our built-in lie-detection skills.

Sign 1: Out of Place Flattery

Beware of the person who compliments you for no apparent reason.

Of course, common sense applies to this statement. For example, an attractive woman shouldn't flag every man as faker who compliments her on her good looks. That isn't out of place, it is due. She looks great, the man notices, and feels like having to say something. End of story. Instead, beware of the person who compliments you on something he didn't even get a chance to find out about you. Someone you just met says "I like your personality" or "you are such a good person" perhaps while putting his hand on your shoulder, and looking you right in the eye – *you just met*! So how could he really know that about you?

We all like positive feedback, whether it's looks, personality, or achievements. Such remarks make us feel good about who we are, and what we do, and encourage us to continue on our path. A compliment means we are in alignment with our fellow man. Fakers seem to know this more than anyone does and they make use of it very frequently. A compliment up front can increase their chance of gaining trust.

There is that little feeling you may get when someone compliments you like that. A split second doubt and feeling a little bit out of place. This is your built-in lie-detector telling you that something is fishy. Nevertheless, you dismiss it when you see the person looking right at you with a reassuring smile. More talking usually follows, which can also prevent you from giving enough thought to the split second doubt which usually registers as a weird feeling. (Feeling out of place)

How could we resist? Who would do such a thing as to fake these emotions, and for what reason? Nah, it can't be. In such case we may rationalize, that he/she must have keen insight or perhaps a sixth sense. *How perceptive! I'm a good person after all. How does he/she know? Oh well ...*

Indeed, you are a good person. The little moment of hesitation when you felt out of place is long gone. It has been successfully overridden by what you see which is the display of positive emotions. A reach with the arm, affection, smile ... *oh my.*

And snap! Just like that. You have fallen under the spell of a faker. This doesn't just apply to women of course. In relation to fakers, "he, him, and his" are interchangeable with "she, her, and hers" and vice versa throughout the book. Depending on what the faker is after this can go on and on for a while building trust. Fakers aren't just after financial gains. They are attracted to all forms of power. Money, position of power, and position of influence are frequently on their radar.

They gravitate towards the powerful and successful. It may not be apparent what the success is that attracts them. It can be as simple as being popular in school. Even more potentially puzzling is when there is absolutely nothing they could be after. Only when we realize that lying is part of their fabric then we can see that they will spread illusion because that is what they do. Their reasons for doing so are subjective and potentially invisible to us.

We can think of them as people with a permanently distorted view about others. "Others", in their case, really mean every single human being other than themselves. No exceptions. To them, woman, men, and children all fall into the category of "others". Therefore, it's unwise to think that they will be selectively fakers to certain people while remaining honest and trustworthy to you. Thinking that you are the one that they would not betray, means their spell worked.

In history, many conquerors used gifts before the attack to an unsuspecting enemy. This is an age-old trick and in the end, once the enemy is conquered, they take back the gifts as well.

Spotting Fakers

Fake flattery and compliments are similar in nature except the gift is not physical but emotional. This subjective gift of fake kindness is the number one tool of fakers. It is a lie and a dangerous one indeed, because you are dealing with a predator in disguise. It doesn't have to be a new connection. You may know someone already who is a faker. In that case, you may consider looking in your memory, and see if you can recall how many times that person mentioned "what great friends" you two are, for no apparent reason. You may be a good friend to the faker and the faker may return small favors. When the time comes for betrayal this won't matter.

Interesting note is that the characteristics of honest people in these matters are the exact opposite of fakers. They mostly (or only) give out compliments when it is due. They tend to acknowledge others for their achievements. Why? Simple. Because they are honest and say what's on their mind. You do something nice for them and they may say they like or love you, but there is a good reason for that. They are showing their true emotions. Therefore it is important not get into false alarms. You pull someone out of a burning car, and they say "I love you man," there is nothing out of place about that.

Fakers rarely give real physical gifts. When they do, it is usually something nonsensical, trivial, useless, or fake.

A faker – who says he is from Hawaii – may bring by a present, for example some pineapple in a plastic container, and tell you that it is special – straight from Hawaii – and how magical that is. He insists, you take a bite and immediately notice a bit of can flavor. If you are not onto fakers yet, then of course you may begin to rationalize, and try to suppress your suspicions. *"It must be the plastic container that had the taste stuck in it. Those darn containers are so hard to wash perfectly. Yes, that must be it, the container."* It is unthinkable that he could be sitting there making this all up. Finally, you dismiss the mental picture, in which he just came out of the nearest grocery store and bought the cheapest canned pineapples, not even a fresh cut one, right before arriving to your place. Illusion is his game and he will make that insignificant gift sound like a royal feast, handpicked just for you.

When we receive information, without knowing it, we tend to file it as black or white, yes or no, this way or that way, etc. The danger is that he will have plenty of lies ready for use and can make them up as necessary. He will tell fanciful stories, demonstrate fake emotions and such, in order to sell the illusions and thereby gain your trust. If you don't call it for what it is, right then and there, at least in your mind, you will likely accept it. Why? It is because our mind doesn't like unconnected dots, which is why sometimes we may feel like "we must know". The faker does get your interest going. Your mind will want to make a decision on the subject of whether this person is trustworthy or not. If you don't call fake on the personality, you will most likely accept it as not fake. Since the mind needs to make some connections, a good alternative, is to realize that you can't tell for sure and thus for now the personality is not known. The urge to connect any new event to something known is always there. This is what our mind does – connects dots, day in and day out. Considering something as "unknown" is a valid connection for our mind. We should acknowledge to ourselves when something is "unknown"" whenever there isn't enough information to go on.

The previous is just one example of a faker selling something off as special. He may follow it up with a little story about just how magical Hawaii is, with a bunch of overblown expressions, and again you question, *why on Earth is he here, if it is so magical there?* However, you don't say it; to do so would be rude. We don't say these thoughts because we are honest, and don't like to be confrontational, so we just go with it. We may even feel dishonorable or tainted from such doubtful thoughts, and that is enough of a reason to dismiss them.

There are many forms of flattery, but the "out of place" part is very important. It takes practice to call fake and trust our judgment. We have ages of missing information. People have lived countless generations without widespread knowledge of these facts. As a result, we tend to find all sorts of justifications, as to how the sign we observed, is an exception. For example, when we meet a new person we may have thoughts like "He grew up in the sixties, it's a hippie thing, and he loves everyone". Fakers can and will pick up a hippie personality in a heartbeat if they see it as a good camouflage. This of course doesn't mean that hippies are fakers. The true ones just simply do love all people. Fakers will use anything and everything they can to disguise their real purpose and feelings.

The signs are accurate in spotting fakers, but only when used properly. If the signs are there, you have spotted a faker. Instead of dismissing your suspicions consider doing your research; and finding out for sure, otherwise the consequences may be regrettable.

Fakers pretend to be friendly and trustworthy. In order to gain trust they routinely convince honest folks that they are friends. In fact, best friends forever through thick and thin. This is still a form of *out of place flattery*, except this is an ongoing continuous one. The honest person is reminded by the faker routinely what great friends they are. He will usually feel the out of place sensation, every time the faker says something in order to elevate the magnitude of their friendship but gets used to it. The reality is that they are *not* such great friends. The honest person's mind knows this and tries to warn with the split second thought, which registers as the out of place feeling. Due to the illusion and smiles, he dismisses (overrides) these thoughts. The irony is that the honest person *can* be counted on as a friend even though he doesn't try to artificially elevate the level of friendship. Meanwhile, the faker who frequently speaks highly of the friendship will eventually betray the honest person when the occasion arises.

The same phenomena can also exist in intimate relationships, but we should examine the signs carefully. Compliments and flattery may indicate ongoing true love, and attraction. There is nothing wrong with this and shouldn't be confused with *out of place flattery*. The out of place feeling or thought must be present *along with* the flattery. Forgetting this and focusing on the flattery only, is useless in detecting fakers.

11

Miklos Zoltan

Fakers constantly work at maintaining their image of trustworthiness. Therefore, if you are associated with a faker already you will see – if you look – that they will be dishing out compliments for no obvious reason. Sometimes – or all the time – this will give you a feeling of being out of place, like being temporarily frozen for a split second, unsure where to put the sugarcoated praise. When you have that feeling while talking to someone, you might just be on to a faker. It's not like we don't know, we are just too decent and unassuming to consider that such emotions are fake. It is easy to even the odds. Once we understand fakers, we can realize that we shouldn't feel ashamed for our doubts. They are the ones who should be ashamed for lying.

Sign 2: Know it All

Most fakers are not exceptionally good at lengthy projects or inventions. Some of them may be very clever and may even strike you as raving genius, as they can dramatize being a genius so well. Fakers can shine especially when it comes to proving someone else's work is incorrect. Their mentality of self-elevation really shows there. It is as a warning sign if they can only shine, when correcting, or discrediting, someone else's work. The favorite challenge of fakers is to prove for others, as well as for themselves, that others are not as good as they are.

Cleverness should not to be confused with honesty, or being trustworthy. Cleverness means that one is able to combine and manipulate quite well. Being clever is like a tool, and just like any other tool, the purpose is up to the user. Since illusion is the game of fakers, they will present all of their skills with exaggeration. If cornered in some field, and we spot that they don't know as much as they claimed, they have the perfect mechanism to cover it up. They will lie even more to save face and create even more illusion. They may switch the subject, and try taking the argument to a different field of knowledge, or science, pretending of course to be an authority in that field as well.

For example a person, who spotted the hole on the faker's knowledge in a particular field, is an honest expert at that field. As the faker switches the subject, the honest person will eventually find himself in a field where he is not an expert and back down, since he is not into pretending. Then the faker walks off with his fake victory. They are very hard to hold onto in such conversations or debate. It's like trying to hold a live fish with your bare hands, or better yet with plastic chopsticks. They are very slippery in this aspect.

The aftermath of such debate is also important to look at. The honest expert may know with great certainty that the person was faking in his field, but some of the illusion of the faker may work even on him and cause self-doubt. "Maybe he is right" – thinks the honest expert. All because the faker is so convincing and seems to know so much. He (the honest person) might even think his vision is narrow because he only knows one field, while the other person appears to know so much more. This is just one example of why we don't call fake. If others were present at the debate, they will be even more likely to accept the fake person to be right, especially if they are not too familiar with the field where the debate started.

Now the honest expert, if one asks about his or her expertise, the answer would go something like this: "I'm fairly good/ok at it" or "I get done what's required" or something plain, sincere and simple.

Let's look at why the faker's "knowing it all," and "knowing it better than anyone," attitude is a concern. Time is finite. To become professional at something it is necessary to spend ample time with that subject. It usually takes some years for each profession. If we look around, and see how long it took real professionals to become so good, we can realize that they learned and practiced for years. It is important not to ignore facts like people who started at an early age. A child, who starts playing with electronics at the age of five instead of playing with action figures, may have a huge head start compared to others in the field of electronics. If he stays on that path, by the age of 16 could be a capable electronics engineer. When studying how people become professionals, it is important to notice such head starts, as well as devotion and overall time spent with the skill in question.

In many fields there is so much knowledge that it becomes necessary to specialize in a certain area. In the field of medical science for example, doctors don't usually do everything from lab test to brain surgery. At some point, they are required to decide in which field of medical science they wish to specialize. A choice is required to become a general physician, surgeon, pediatrician, dermatologist, etc. Some of these categories get further specialized. For example, a surgeon might specialize in heart surgery while another focuses on the brain. Can anyone be an expert in all of these fields, all at once? Not likely. After the initial learning that is required to become an expert, it is also necessary to keep up with new technologies and discoveries. Anyone who claims to be an expert in it *all* is exaggerating, for the simple reason that they have a finite time during their days, and it takes daily practice and study to be a good expert in any complex field.

If we leave any of our known fields behind even for a year, we may fall out of practice. If we come back to that field, a few years later, chances are we may not even recognize some of the equipment used. Weeks or even months of catching up might be necessary. Electronics and computer science are similar in this regard. Someone may claim to be a professional in all areas of electronics, or computer science, but that will be a lie. New technology, new chips, and new software emerge on daily bases, affecting anything from cell phones and home computers, to medical equipment. This does not mean that we cannot run into someone in their 40's or 50's, who mastered several different fields already. Such experts do exist, who have practiced so much, that it may seem like they really know it all, but people like that tend to be humble. People, who possess true knowledge, tend to be honest, and don't inflate their abilities. If anything, they tend to play it down. It is easy to recognize such people, as they will tell facts, and when we ask them about something that is not in their field, they don't pretend to know. If someone whom you have not seen for five years shows up, and says he is now an expert in many new fields, such as photography, networking, and computer science, you may wonder how much can one really learn in five years.

We tend not to question boasts about the knowledge of others, as we assume that the person presenting these facts is not lying. Why? We base our view of others on ourselves. Therefore, we assume (believe) that the other person is also honest. This assumption isn't always true.

15

Our minds are capable of taking thousands of factors into consideration and spit out the results immediately. Spotting fakers does work automatically for many of us already, except we tend to reject such thoughts, for the simple reason, that we have missing reality (information) about fakers. Once we understand and accept the underlying logic, and gather the facts of reality about fakers, our mind will do the rest with little effort.

If you are an honest expert in a field and find someone playing hide-and-seek with you in a debate, consider what is real. If the statements seems irrational they probably are, and now we know what type of personality results in such overstatements routinely. In a hide and seek debate, which may have started as a casual conversation, you will feel like there is no way to get your point across because you are dealing with someone who knows it all. Every time you try to steer the conversation back onto the original subject, the faker's response is to change subject, and avoid the original one. He knows he is bluffing and you merely suspect it. He will stay away from the original subject at all cost. He might even blow up, or take it to some personal arena, perhaps an insult. Trying to get a faker to get back to the original subject is futile. You may feel like you are playing whack-a-mole. You make one point then they pop up with a new quiz.

Since in case of a faker, you are dealing with the *illusion* of experience, instead of real experience in whatever field they claim expertise, it is relatively easy to find the holes in their knowledge. If you pay attention to what they produce, cross check and look for errors, you will find proof. The trick is to accept the proof. It's not necessary to justify your findings. If they claimed to be an authority, then there shouldn't be frequent errors in what they do, right?

Those who profess to be absolute experts and yet make gross mistakes in their field are dishonest about their skills. This is an indication that they have serious missing information going all the way back to the principle and basic knowledge of the field.

Sign 3: Bearer of Bad News

Being critical of others, and making frequent negative remarks about people behind their backs, is a sure sign of a faker. Of course, it is not necessary to flag everyone because of occasional bad news or banter. We are looking for a pattern, where mostly (or only) bad news seems to get the attention of the person – enough to share it. We should look to see how factual the report is.

If you notice frequent hearsay and/or gossip from a person, in the form of "Joe said Jay is not to be trusted, because of such and such ..." then you are possibly dealing with a faker. Verifying each accusation is crucial. Indirectness is a convenient form of illusion therefore; fakers tend to hide behind other people, and often make up, or distort news for the worse. Generally, they avoid saying anything negative directly to the subject. When they do, it's illusive and filled with ambiguous statements. They are masters at using language with hidden downgrading meanings. They tend to relay useful and good news only when they can use it to achieve their goals. It seems almost as if it pains them to share some good news. Bad news is the favorite of fakers; especially if it is about someone that is in their way, or someone they wish to control and have advantage over.

The reason behind this behavior is that they harbor covert hostile intentions. Some of them may just be indifferent towards the casualties of their actions but from the victim's perspective that doesn't matter. When caught they will say they didn't mean harm, but it's not recommended to believe that coming from someone who pretends all their life. Their smiles, pretenses, and compliments hide their true intentions. They know they can trick others, but they also know that it is not that easy. They feel urged to keep up the trustworthy image, so they hardly ever show signs of hostility directly. Saying something, negative or confrontational, about the person they are talking to is not usual, unless they are cornered. If they did say negative remarks directly, that would work against the illusion they have worked so much to establish. Instead, they will share bad news, or speak less of others indirectly, behind their backs. It's a good idea to keep an eye on their conduct with other people. What you may find is that he/she will badmouth Joe to Jane for example. Then the next day he/she will badmouth Jane to Joe, behind their backs of course. If you spot this, ask yourself

"What would prevent him/her from badmouthing me behind my back?"

You shouldn't delude yourself with the fact that you are honest, and have always done well by him/her. From a faker's viewpoint, even a saint is just another faker. They will do whatever they planned, no matter how honest you are. You shouldn't except them to be fair or just.

If you catch them in the act and ask them about it, you'll only get more lies, denial, or "Oh, I did not mean it that way." When busted, they will quickly back down and pretend to be a saint. Their reaction will be something like "Oh, it was a misunderstanding", "That's not what I meant", or "I'm just trying to be helpful". If necessary, they will resort to crying. They talk about how unjust the accusation is, when they have such good intentions.

Let's look at an example in a meeting. Honest Engineer: goes on and on, talking about a plan B solution he introduced years ago, but no one listened. Due to the pressure of new challenges, he says, "Implementing plan B is inevitable now". He adds, "We are running into more and more problems with the old system. The new solution will take more time, but it is inevitable. It will pay for itself in more than one way in the long run."

Faker manager: "Ok, I understand all that, but let's try to be realistic here, and stay on the ground".

Spotting Fakers

Honest Engineer: "Ok, ok, I'll patch the old system, and add one more feature".

This may not look so offensive at first, but if we look at it in detail, the "Faker manager" just insinuated that the engineer has his head in the clouds, whereas the engineer is the one who knows the system.

If he knew about fakers already, he might notice the insinuation and say this instead:

Honest Engineer: "Are you saying I'm unrealistic? This is a present danger and you are implying that I'm not thinking clearly."

Faker manager will have a response like:

"That's not what I meant …" (But he did!)

The interesting part is the aftermath, because the faker will make sure, one way or another that the engineer loses a bit of his image. If the engineer just agrees, then the faker will get away with calling the engineer somewhat crazy. Such subtle remarks are an attack in disguise. One of these may not make much difference, but fakers use these regularly. This makes these remarks stick after a while. After the meeting, the Faker manager may make some remarks behind the engineers back like:

"Engineers, always with their heads in the clouds, right?" or "Engineers ... they are always so sensitive", or "Such Prima donnas".

Either way the faker will put in at least one negative remark about the engineer. He has to appear to be in control and having to tell the engineer how to think. The illusion the faker will put forth, is that he could do what the engineer does any time, but he is just too important for such work. The illusion the faker is building is that without him leading, the engineer would be lost. Besides, he can see the big picture only, so he can't degrade himself with petty stuff, like engineering.

Fakers use hidden condescending expressions frequently, in a safe ambiguous way. By being ambiguous and indirect, they leave a convenient escape route. Sometimes they slip up in the heat of a busy meeting or debate and spill their views directly but this is infrequent and we tend to dismiss it. We may consider it as an unimportant mistake but in case of a faker, this is the only time you see the true person. They continuously have to deal with the overhead caused by their lies. This gives them a disadvantage, and thus they may be agitated in meetings, for no apparent reason, especially when their superior is present. The fear of slipping up is at large at these gatherings. This is where they really have to pay attention, and this can drain them. They may even stutter at meetings because of the pressure of their own lies. Stuttering or being agitated at a meeting alone, doesn't automatically make one a faker. The signs are required to be present in combination and finding a pattern is as important as ever.

When a faker has an issue with someone, who stands in the way of getting what they want, they can get very hostile with that person verbally, behind their backs.

Due to their constant illusions, lies and hostile intentions it never feels safe for them to talk about what they are doing, did, or intend to do. Instead of saying what they feel like doing they make it indirect and say how someone else should do this or that. If one of them becomes a dictator, or leader of a "cause", he may end up doing violent things to others who get in the way. The monsters of our global history have done such things, and this is a fact. How could they do such things? Simple: "others *do not* matter". This is their rule #1, or #1 rule. This is the basic rule of fakers concerning all others around them. Later chapters will explain how and why they accept this rule.

This is a hidden rule, which they hardly ever reveal to anyone. Often times they aren't even aware that their actions towards others are governed by this rule. If they make it to power, they can present a grave danger to society. When they do, they will force others to do their dirty work as well. This helps them to keep their distorted life views and rules intact. When others do bad things, it strengthens their view that others are no good either.

Spotting Fakers

It can be very useful to observe and think about what people say about others. They are giving us insight into their personal thinking. The subject someone brings up to us, regardless of how irrelevant or unrelated it seems, is something that person has attention on. Fakers camouflage their own intentions by presenting them indirectly as if it was someone else's. It isn't.

Receiving attention from a faker is usually bad news. They only tend to have attention on those they plan to swindle and those who get in their way. Due to rule #1, we can be that person, no matter what great relation we think we have with them.

Sign 4: Not Answering the Question

Deception is a familiar strategy used in wars throughout history. Countless leaders used this as a method of employing forces in combat to overthrow the enemy.

Rational people, who have no hidden agendas, have no problem answering simple factual questions with virtually zero delay. When questioning fakers about their actions, they frequently avoid a direct answer, take too long to answer, or their answers are ambiguous. The reason for this is that fakers pretend, exaggerate, or lie very frequently. Because of rule #1 (Others do *not* matter), they feign concern and sensitivity. Their true motive is to gain an advantageous position and win confidence by trickery, illusions, and deceit. They are careful not to contradict previous allegations. Caught in a lie would blow their cover, and they know it.

Direct answers may conflict with a previous lie or illusion they worked so hard to build. Before they answer a direct question, they must double check that they are sticking to the previous lies and illusions. The human mind is very fast and impressive but fortunately, all the lies and illusions a faker spreads are too much to handle even for such outstanding mental powers. Some additional delay in response is inevitable, relative to the question. Some questions require time to think before one can answer, but simple direct questions like "What did you have for breakfast?" do not require much to answer. Even with such simple question, a compulsive liar who doesn't want to be caught, has to think first. They might have thoughts like "Did I tell someone already that I skipped breakfast today?" They might have used such lie already to gain sympathy during an explanation as to why they were late from work. A faker might hesitate or even avoid answering such

23

trivial questions. They also use small lies throughout the day to increase the affinity of others, towards themselves. The more someone likes them the more they will be trusted.

Another example is that he may have already told the receptionist that he skipped breakfast, because he is on a diet also. Just to be able to strike up a conversation. In reality, he had a large breakfast. They know they can gain affection from us by appearing like-minded. They will pretend to have similar interests and lie accordingly. Each little lie, which makes them appear similar to us, helps them one-step closer to our trust. All these lies require a lot of crosschecking in their minds.

The key is to trust our judgment, and notice when someone seems to answer simple and direct questions with unreasonable delay. It's normal for people to be consistently slower than others. Some people just think more before they respond to questions in general, but do so consistently. Fakers on the other hand will show inconsistency. Typically, factual questions about their actions will have significantly more or infinite delay while they will have no problem giving split second negative responses about others. Since fakers lie as needed, the lies different people hear may not match. We can expose them when we notice this. Of course, they can always make up more lies to cover up the previous ones.

It takes time for them to run a consistency check on their fabrications. In order to cover up the delay sometimes they just "uh ..., ah ... err ..." their way out of it. The clever ones will start answering something else, or answer another aspect of the question that is irrelevant. Chances are that the original question will require repeating. This may give them enough time to respond with what may appear to be a reasonable answer. They may choose not to answer at all, even if we repeat the question several times. This is a telltale sign.

Spotting Fakers

One or more of the following reasons cause this sign:

(1) They simply don't know the answer, but they can't say that. It would expose them as a fraud about their authority.

(2) They won't answer because the answer would contradict a previous lie, they told to one or more of the listeners present.

(3) They're so overwhelmed by their own lies they believe that no matter what response they give, they'll be caught in a lie.

Backed into a corner, fakers do one of the following:

- Change the subject
- Use verbal attacks
- Pretend to be offended
- Leave

Changing the subject is a simple tactic to avoid the subject in question. They do this because the response may potentially expose a previous lie or illusion. If it works, fakers will never return to the subject voluntarily.

Verbal attacks take the position "the best defense is a good offense". Usually this can successfully deceive the other party, by forcing the attention to other subjects. Fakers will resort to baseless accusations and try to make them stick, through their excellent acting skills. Upsetting the other party in such way allows the faker to keep the attention away from the original question sometimes permanently. (This is a violent form of changing the subject)

Pretending to be offended is a powerful tool they use to get sympathy and appear to be caring and sensitive. They may even resort to tears (many of them can do that at will), and pretend to be shocked at the so-called offense. They will present imaginary events as real or bring up and distort past and recent events to gain sympathy. They will pretend to be the victim. They can act agitated, loud, and convincing and appear to be offended to the core.

Leaving is a convincing technique fakers use to turn the tables, thereby making the other party feel guilty and difficult. Leaving combined with playing the victim, guarantees that the other party will look like a bully. Bystanders may think he (other party) is the insensitive one. This can taint his image and reputation permanently. He will have a hard time shedding that image, especially when the faker manages to make him feel bad. He may end up thinking he is indeed insensitive because he doesn't know it's an act. This is a powerful illusion.

Observing their manipulative tactics is valuable for the rest of us. Consider keeping an eye out for those who are repeatedly slow, dodgy, misleading, or become emotional when questioned. Those who repeatedly *omit* coherent and reasonable answers to simple direct questions are most assuredly faker suspects. Of course, other motives may exist, but keep in mind that honest people usually want to clear misunderstandings and answer direct questions.

The keys to look for are "repeatedly" and "unreasonable delay". We shouldn't flag someone as a faker for taking time occasionally, before giving an answer.

We should only consider the delays related to direct and factual questions.

Complex, puzzling, challenging, educational, and scientific questions are not useful for spot fakers. Taking time is natural for anyone with such questions.

Sing 5: False Accusation

It is self-evident that those who are out to get others will be expecting the same in return. This is because they base their view of others on themselves.

As a result, fakers dish out many accusations. They are up to no good and think that everyone else is also. When the one being accused is an honest person, the accusation is usually false. Fakers do this because they have a constant mistrust of others. Therefore, telltale sign #5 is people who frequently make false accusations about the people around them. The falseness isn't always obvious because we might be missing some details. However, the frequency of negative accusations is always obvious.

A faker never misses a good opportunity to take a jab at someone when a good chance arises. Any time is a good time, if he feels he can get away with it. Often, he is quick to point out other people's imaginary faults, shortcomings, or character flaws. (Imaginary means the faker imagined it) "You are so hard to deal with", "You never listen", "You make too many mistakes", or "You are so mean" so on so forth.

When the sentence begins with absolutes, it is a ploy used to belittle a person or group, thus making the faker feel or appear powerful:

"You are always so ..., you are this you are that, he can be so ..., she is this, they are this and that ..."

The focus is never on the faker. Instead, he dishes out negative remarks regularly about others, but never admits his mistakes or participation in less than ideal situations. The blame falls on someone or something else. The accusation is usually indirect and

27

about someone else who is not present. When upset or cornered he will do this directly also.

Unfortunately, some of these remarks stick to the honest person, causing mental anguish, since he will consider the accusation as truth. He reasons, that there must be some truth to it even if partially. The negative remark is usually unmerited, when it comes from a faker. The honest person may suffer because of evaluating whether the negative remark is true or not. If this person is self-conscious (and honest people are), every effort will be made to be better. The problem is that he is not the one who needs to get better.

For example, a faker may say to Joe: "You are so selfish." Joe, not knowing about fakers, may end up falsely accusing himself of being selfish. Meanwhile, the truth is that the faker is the selfish one. Fakers think everyone else is selfish like they are; therefore, they frequently accuse others of being selfish, cunning, scheming, or dishonest.

These types of remarks don't always stick. It all depends on the circumstances and the person who receives the remark.

This activity of fakers also adds to the overall damage they cause. Mistrust is their constant viewpoint.

To sum it up, these are the things to keep an eye on: how people deal with other people, and if they speak negatively about others behind their back. Profuse accusations, and chronic mistrust, are sure characteristics of fakers. Whenever we accept a false accusation to be true without due verification we grant power to the lie presented. An intentional false accusation is based on one or more lies. The liar who presents it gains power with every false accusation that is uncontested. Fakers are very good at selling such accusations using their fine acting skills and custom tailored stories. When we don't know they are lying we tend to accept at least part of the accusation.

Sign 6: Feeling Entitled

The mentality that "others do *not* matter" automatically elevates the one who embraces this line of thinking. *"Only I matter."* This characteristic of the faker personality is why it's so easy for them to rip-off others, time after time, without feeling guilty. The result of this behavior is a bottomless pit of never-ending greed and a sense of entitlement. In their viewpoint, they are only receiving what is due.

Their existence proves just how immorally they are able to manipulate circumstances by this guiding rule. This is a potentially dangerous mentality.

Countless movies and books are available about Hitler's Third Reich during World War II. It seems obvious that he was operating with Rule #1. Considering the past, one thing is certain. Hitler's path was full of lies and broken promises. He had no difficulty wiping out millions of people, including his own, if they got in his way. Obviously, the ideology he persuaded his followers to believe was in essence a lie, although they carried out his heinous orders to the letter. His legacy is a grim one, but if monitoring fakers could prevent the rise of dictators like him, it would be worth the effort.

Wouldn't it?

People like them have the ability to camouflage their true attitude and hide rule #1.That is not to say that all fakers possess the harmful and highly negative personalities as a Hitler – but can we afford to be complacent? We like to think that grim events, such as an immoral dictatorship, could never happen again, but the potential for such monsters are still among us. Fortunately, we can do something about it *now*.

Feeling entitled without proper exchange is very common among fakers. We tend not to question the facts of exchange. We can even ignore the facts and make up our own rationalization in favor of the fakers, as we don't naturally assume unfairness and lies. In order to spot a faker using this sign it is important to remain rational and look for the facts of exchange. There is nothing wrong with someone feeling and pursuing entitlement if they have contributed fair exchange. The logic and ego of a faker can have him truly convinced that he deserves to run a whole company just because he worked as receptionist there for a month. The remarks they make may be indirect but if we pay close attention, we can spot the enormous ego. (Only I matter) It can be just a quick remark about how they should be the leader or in charge of something. We tend to dismiss it as a joke and laugh it off with them. They aren't joking.

Sign 7: Reporter of Conflict

This example shows a conflict between an honest man and a faker. Chances are of course that the faker was trying to exploit him. For this example, the story is that the honest man ordered a service from a faker, unaware of whom he is dealing with. Honest man bank wires 20% of the money in advance to the faker. After arriving, the faker says he can't deliver the service as previously agreed.

This could be due to any number of fundamental disagreements, but in this case the faker said he could do something, which turned out he couldn't do exactly as promised. They discuss it back and forth as to why not, until the honest man just gives up and tells the faker to go back home and keep the money for his travel expenses, and trouble. The honest man assumes that the faker was not tricking, and didn't know ahead of time that he can't deliver.

In reality, the faker was exaggerating his skills, which is why he could not deliver the service. Such a simple lie means nothing to a faker.

Even in such a clean cut case, where the faker is at fault by promising something he can't deliver, and getting away with money he didn't earn, the faker will do this:

He will badmouth the man who hired him.

Why does he do that?

Because a preemptive attack, is better than allowing others to find out about the ordeal from someone else. Fakers expect the other party to do what they would do. The belief of the faker is that the man who ordered the service is also a faker, and thus cannot be trusted. What if the honest man, tells everyone about the deal gone sour? Eventually, the true details will surface. The faker doesn't think for a second that he will be off the hook without repercussions. To avoid this, the faker's strategy is to spread lies about how he could not work with him: "he's impossible, he's crazy, or he wants the impossible done …" etc.

The moral of this, is that when we first hear a story of a deal gone sour, we should consider paying special attention to the person who complained first. When deals go bad, fakers plan preemptive attacks guaranteed. (Deals usually go bad whenever fakers are involved) Not only do they rip-off people, they also immediately badmouth the person to damage their personal and professional reputation. After all, when people finally hear about it, it will seem like a conflict in which people can take sides. Once people take sides, the faker further manipulates with his magic touch of illusions.

One unfortunate factor here that benefits fakers is that we tend to believe the story from the first storyteller. The root cause goes back to how we store, learn, link, and know things.

When we first hear the news of an ordeal, our mind has to file something about it. The mind has to make some connections with this new information. If we don't expressly tell ourselves not to believe the story without verification then we are likely to believe it and pick a side. *Which* side? The faker's side and this is why: the presentation is detailed, and it contains great emotions and convincing acting. This works in favor of the faker who complains first. When we are unaware of them, we tend to lean to their side. Using their acting skills, they can tell the story with so much detail that we naturally identify with their upset and unfair treatment. They can turn on tears at will because they make themselves believe this was unfair. After all, the faker (since he feels entitled) thinks he should have received full payment for just talking to the customer. This may sound comic, but this is how they think all because of rule #1. In case of a conflict, there are always two sides at least, which is why it is wise to remain neutral until we have all the necessary information verified.

Spotting Fakers

It is a good practice to think like a judge and do what they do: listen to both sides, look for evidence, and avoid quick judgments. The chances of being fair are better if you side with the one who didn't complain. This is because honest people don't like to share bad news. Therefore, they tend to do that only when necessary.

Sign 8: Faker vs. Faker

When two (or more) fakers get into a conflict or debate things can flare up. In a debate, you may see them really get at each other viciously. They both operate using the same principle. (Rule #1) They will change subjects rapidly, each trying to upstage the other. When the subject changes, the other faker responds as an expert of the new subject also. It may seem like a rooster fight. Important to notice how they position themselves as unquestionable authorities while they tear through more fields of knowledge than anyone could possibly gather even in two lifetimes. If you have any doubts about their skills in the subjects they tackle, don't ignore your instincts, you might just be onto not one, but two fakers. Of course, this is not to be confused with a heated but respectful debate. You should make a choice. If instead of reason, frequent subject changes, sketchy reasoning, insults, mockery, or personal attacks are used, take it as a faker warning.

The intention of fakers is to make the other party feel inferior and upstaged. To achieve such effect they manipulate the tone of their voice, how they phrase sentences, facial expressions, and body language. They might yell, scream, jump, or even fake a seizure. There is almost no limit to what they are willing to mimic to get what they are after.

The reason should be obvious by now. Most of the time fakers are hard at work keeping their illusions alive. They can't be caught upstaged, no matter what. Consider being extra cautious if you have two guests who go at each other in such a show down. They might just be considering you as a mark – and you might be part of their turf wars.

Miklos Zoltan

When the motives for the conflict are not apparent, see if ego and projected self-image offers a good explanation. If you watch closely to see who is dictating the change of subjects, who is dishing out accusations and who is avoiding answering questions you'll have no problem spotting the faker or fakers. All is required of us is to kick back, and listen to them go at each other.

Sign 9: Leverage Seekers

Imagine that after ten years of outstanding service, where you have directly contributed a great deal to the company's growth, you get a new manager. Shortly after that, during a meeting, he walks up to you as you sit on a chair, leans forward, puts his hand on your shoulder, looks you right in the eye, and says this out loud in front of others:

"We really need you to be a super engineer now, on this project."

What do you do? You might have something like this on your mind: "Well ... even though I don't brag about it, I thought of myself as a super engineer already based on my achievements."

If you don't say anything he just publicly, in a sneaky indirect way, undermined your outstanding engineer status. Your status may have been unspoken before that, but you could tell that others thought highly of you, even if they didn't directly say it. Now there is someone voicing the opposite. Once said it can taint your reputation. Why? Because your colleagues are also honest and they don't assume the new manager is up to no good. The words of the new faker boss will damage your image. There are ways to deal with such situation, but ignoring it isn't a good idea. You just got a faker as a new manager and he is seeking leverage over you. Say you have been putting in 60-70 hours a week already, for all those years and have been with the company from the beginning. He wants to make you feel like you must do even more. He is new, he needs to worry about his own performance and statistics, and he is not going to give a damn about you having to steal even more time away from your family. He needs you to provide more for the company, so he can look good. Because of rule #1 you, your family, and just about every employee and their families as well, can pay for the extra burden, just so he can appear better to his superiors. That is if you let him.

If he is a faker, you're in trouble. If you comply, you pay the price and he'll want more and more from you and will never be satisfied. He may even have you question your worth. We can be vulnerable to such suggestions. As mentioned before we tend to assume the faker speaks the truth. If he succeeds, he will get you to work more to the degree he demolishes your self-image. The more you feel your production questioned, the more you will do to make up for it. If the faker is a good actor (and they usually are), you will find yourself putting your own productivity under the microscope instead of his. In a situation like this, people can even end up with thoughts like: "He must be right after all I am not getting younger and must be less productive than when I was younger". Even if in reality, he is the most seasoned core employee of his trade. Fakers will intentionally make us feel less worthy to get more out of us.

If you stand up to him, you'll be the black sheep. If he can't have leverage, you will be a splinter in his eyes. Either way you will not win through him. You will eventually be best off being transferred, getting him handled by exposing him, or look for a new job.

The leverage-seeking faker may pretend to seek your friendship, advice, and put forth the illusion of respect.

Another example: Imagine that your new manager invites you and a few other employees to dinner after a meeting. Beers are on him, he insists, so you have one. Then he insists again that you have another. You are a one-beer kind of person and naturally refuse. He reminds you that he is driving and offers to drop you off at your hotel. "I left my car at the company" – you say. Then he says, "It's all right, I'll pick you up from the hotel in the morning, I have business in that area anyway". He insists so much that you give in to the second beer just to be polite. He appears friendly, talks about long-term goals, family values, and even his hobbies. He gets you to open up by asking your advice on a few subjects.

Spotting Fakers

When the intentionally plotted two beers loosen you up you confess a few of your concerns with the upper management, and your dismay that no one is listening to your professional advice. You also tell him that you don't know how much longer you're willing to stay onboard if they don't change.

After dinner, he gives you a ride as promised, and everything seems normal. You don't realize that you've just been had by your new leverage seeking manager. He has all he needs for making you look like an unstable link in the chain, who could quit any day. He uses this to make you work harder.

Now you feel like you need to prove something. He has you at his mercy, and there isn't much you can do about it. You can bet that he will also use this information behind your back, to diminish your image.

Leverage-seekers zoom in on the slightest mistakes you ever made, including any verbal slipup so they can continue to be in control. They might even resort to blackmail. He won't report any of your outstanding achievements. If he does he will distort it just enough to make you look less able. You might want to watch out for others also. If you see someone trying to get leverage over your colleagues, friends or family, it would be reckless to think they wouldn't do the same to you. Just simply speaking up about it can make all the difference.

Since we are living in a society abundant with fakers, it is best to watch whom we trust with our opinions. Honest concerns can and will be twisted by fakers to use against us mercilessly.

Sign 10: Knowledge Bullies

Fakers have a natural tendency to try to prevent this type of information from reaching the public, as soon as they realize that their way of living may be at risk. If this information should become widespread, they can kiss their scheming ways goodbye. That's because if overall awareness increases, it will be easier to spot them.

It is possible that this book will cause some commotion. Fakers will not like this message. Therefore, we should pay close attention to those who try to belittle this knowledge, or any other knowledge exposing them. They will do that to prevent us from spotting them. Fakers use opinions, and emotionally charged subjects, instead of facts. Replacing logic with opinions and personal insults means someone is trying to steer us in the wrong direction. We are all free to think as we wish and look at the facts presented and think about them for ourselves.

Most people like peace and would prefer our society to be based fairness. Fakers adversely affect the rest of us. They are a major contributor to the temporary ruins of our society. Suppressing knowledge is a usual trait of fakers. We all possess the ability to tell truth from lies. Knowledge is precious and no one should have the power to tell us or influence us about what we should think, know, and study.

Many books introduce theories and excessive volumes of knowledge, which require some time to digest. There are many opinion leaders, and critics, who are not willing to go through that experience. The greater the effort it takes to understand something, the less likely it is that a general critic will have anything useful to say about such subject.

Fakers, who pose as authorities in any given area, feel threatened by new ideas, as they see their illusion of authority in danger. When this makes them feel cornered, they will attack by any means necessary. They will do what they usually do: change the subject, and using opinion leading and emotions, try to destroy the foundation of the individual who poses a threat to their territory. Since we assume they speak the truth some of the slander sticks. In our current societies, there is no shortage of fakers, who claim to know it all, and pose as the final indisputable authority. They essentially bully other scientists in their claimed arena of knowledge. This is the number one reason why radically new ideas meet with such resistance, today as well as in the past.

In the absence of reason, a self-biased critic (or territorial faker) can also result to scare tactics such as suggesting that simply reading something can mess with our mind. They will do whatever it takes to prevent others from looking at the knowledge that could expose them.

Sign 11: The Spiel

Since fakers spend their lives creating illusions, they can get very creative at convincing others. One of their tools is to keep making the story up as long as necessary. The longer they talk or write the more likely that the quantity of their message will override the listener's initial suspicions. They may end up creating lengthy reports, novels, or even books based on the fabrications of their mind. They do this in an effort to cover up their lies. When we hear a story or pitch like that, we tend to have a moment of doubt at first. That is when our built-in lie-detector is trying to tell us that something is not adding up. As the faker goes on, it is only natural that we override this feeling with various thoughts. The most common reason is when we ask ourselves "Who would go into such trouble of making all this up?" If we wouldn't do it then it seems illogical that someone else would.

Fakers can have seemingly perfect logic as they weave their stories except for the thing they are hiding. The lie or false-information, which they intend to hide, is what makes their presentation false. Complete books exist which are based on lies. When someone sells a fake story, those who believe it can get upset or hurt.

There was a story circulating on the Internet, which makes a great example. It was about an alien encounter with the greatest attention to details. The scenery and how the alien was captured were described as vividly as if someone was there observing the events. The characters and their dialogue were seamless and sensible. The story spared no details about the looks, origins, habits, abilities, and intentions of the alien. All characters received such detail as if someone filmed the whole event then later used it to create a transcript. Everything seemed perfectly believable and natural except for one thing – the story claimed the alien spacecraft crashed because of lightning. It is easy to skip over that one fact, as the story is so fascinating. By the time the readers get to that part, they are likely to want to hear the rest anyway. The storyteller was great and authentic.

It is estimated that each airplane in the US commercial fleet is hit by lightning at least once a year. Considering that, it would seem very hard to believe that a spacecraft, which was able to endure the harsh environment of space, would crash from a lightning strike. The same type of lightning our airplanes can take routinely without a hiccup. If an alien race managed to fly one of their members here, their technology would be superior to ours. (We haven't even landed a man on Mars yet) Of course, one could speculate at that point that perhaps we have some real heavy weather conditions here and perhaps our atmosphere is so violent that the aliens have not yet experienced such and thus came unprepared. Indeed our weather can be very harsh at times but it only takes about 5 minutes of research to realize that compared to space travel our atmosphere is like a walk in the park.

Space as we know has plenty of potentially deadly radiation, high charged particles, asteroids, and more. Additionally the story later revealed that other members of this alien society have been visiting our planet routinely for scientific observations purposes. There goes that idea of them being unprepared. If they are frequent visitors here, they had plenty of opportunity to experience our atmospheric conditions. The story also describes the ease at which these aliens can and have been travelling through space and monitoring other life forms.

They can travel near the speed of light, hopping through "worm-holes" and such. Yet they come to Earth and get downed by lightning. Really?

If we insisted on believing the story then there are a few more options. Perhaps this particular alien was somehow unlucky and had a malfunction of some sort on the ship. Perhaps he forgot to take his ride in for the "every two billion light year service".

Spotting Fakers

Except in that case, among all the meticulous details of the story we would expect to hear more than a one-liner about the crash.

The dialogue between the alien and our "representatives" was described word for word but no mention of the circumstances of the accident? But hey it's a freaking alien encounter! Who would think straight in such a situation right? Well, according to this story everyone involved was. The details of the questions to the alien were practically down to the color of the toothbrush the alien was using back home.

This story shouldn't be taken as a mockery of UFO stories and alien encounters. For all we know some of them *are* real. The point is to be able to spot the one sentence, which makes the story crumble. The spiel may go on and on until our ears fall of, but when we pay close attention and trust our judgment we can call the fake on the one line that doesn't add up. When fakers want to convince someone they will do whatever it takes. Their motive can be as simple as getting attention and popularity. What's more likely?

- The alien story is real, and despite the virtually zero odds, the spaceship crashed because of lightning.

- Or, the writer had a vivid imagination and excellent story telling skills, but lacked the technical background and therefore couldn't come up with a solid lie surrounding the crash.

In this case, the writer did seem to avoid technical issues, which in combination with the unlikely scenario of the crash just furthered suspicion. As science fiction, it would have been great.

The problem is that when such a story is presented as a real and trustworthy report, those who don't discover the lie will be living with that lie. This can create confusion on the subject. In addition, it can be disappointing to find out it was a lie. People's feelings and hopes can be hurt. That however, is of no concern to fakers. If they see any possible benefit from making up such stories, they will try to sell it. They can give the best writers a run for their money.

Miklos Zoltan

As usual, we should be able to trust our instincts and not let the lie out of sight even if the rest of the story appears very convincing. Fakers also use such story telling skills to diminish the reputation of others. They can make up a lie about someone who is in their way and follow it up with as much sensible filler story as needed. It can appear very believable especially when they present it in person with their excellent acting skills, which would even leave some Hollywood actors envious. Fakers don't need fresh-cut onions or special effects to cry.

How to Use the Signs

Any one of these signs alone can accurately inform you of the presence of a potential faker if you find it to be present consistently. The key is to not only notice what one does to us, but also observe what one does to others.

If you find someone who gives you an *out of place flattery* for instance, you shouldn't wait until you see the same person says it to you again. When you see that person saying *out of place flattery* to anyone else, then you've observed a behavior pattern. You may sense the out of place feeling, even if the target of the flattery isn't you. Consider all of these signs as your tool for detection. Instead of waiting to see if one particular sign shows up, keep an overall summary. All the signs count. Fast or rushed conclusions related to the signs are seldom accurate. For good accuracy, it is best to observe and collect information over time. If you suspect someone is a faker, do your research. If indeed the person seems like a faker, and it is in your position to make the call, consider it carefully before you entrust them with power.

Fakers can remain dormant for decades and would have you convinced that they are your friend. The sure way is to watch for the signs and when you find someone who possesses the signs, keep a close watch on them. Fakers can't help themselves. The signs will be there.

Another aspect to consider is that if you spot someone as a faker once, it is not wise to assume that a few years later he will be trustworthy. Their #1 rule isn't likely to change. If it did change, you would hear about their revelation before anything else. If it *seems* like they changed for the better, chances are, they just got better at hiding their true selves. It is best to retest them from scratch again looking for the signs. The court of law is unforgiving to liars for good reasons, and the same reasons apply to all conduct that affects lives.

The future: if this information becomes public, eventually some fakers will read this book and try to use this information to be less obvious. They will be using this knowledge to get even better at faking. This may reduce the obvious signs a bit, but they can't fix it all. Their actions depend on illusion. If they don't spread lies and illusions, then they might be fakers inside, and still think with rule #1 but they will be unable to manipulate and betray us.

The danger of letting fakers go unchecked is great. Here is a recent example: in 2008 President Bush publicly admitted his greatest regret during his presidency. The Iraq war, which was based on flawed intelligence. That mistake cost us trillions of dollars. That is over 1,000,000,000,000 dollars! All because of a false intelligence report. What do false intelligence reports boil down to? False-information such as lies and mistakes due to assumptions. It is hard to tell which type of false-information without proper research and insight. Honest mistake or intentionally falsified by faker(s) or both? Perhaps assumption caused by omitted information? While proving it either way is not in the scope of our subjects, one thing is certain. We have an increasing number of fakers drawn to our government. Would a faker falsify information if they thought they could gain from that? If they thought they could get away with it, they would. Will they care about the consequences to others? This should be a rhetorical question by now. Of course they don't, since others do *not* matter to them.

Fakers thriving unchecked in society and in our government, doesn't look like a winning formula. We can't afford mistakes like this. We all pay the price one way or another.

The effort of humanity to cure the problems caused by fakers goes as far back as history. Throwing people into jail after a crime, raging war on an oppressor, are all symptomatic treatments. It's time to cure the root cause, and stop fakers from continuously messing things up. They contribute greatly or directly cause the collapse of all that we build. Ultimately, it would be best if everyone knew just enough about them so that we can prevent them from abusing the powers they receive. Honest people are genuinely caring by nature, and might be unassuming but they are the ones who truly propel our society forward. Honest people take the abuse until they can take no more. We are not out of the deep end yet, not by a long shot. If we don't take responsibility for fakers, things will only get worse – again. No law, regulation, or government can ever replace the necessity of every one of us having the ability to spot fakers, and know not to trust them – especially with position of power.

It may seem like a feasible idea that if we constantly monitor someone's action we can trust him or her with anything. We may think that we can revoke the power at will. Since we are overseeing how the power is used, we think things will be just fine.

It is not possible to oversee every single aspect of the operations of someone. Say we decide to give the power of a manager position to a faker, out of human charity, thinking that we can make it work by keeping a close eye on him. The problem is that we can't be there all the time. When we are not looking, the faker will manifest regrettable things. He may be working to rip-off the company, or he may be working his way towards freedom from our protective and caring attention. He will do everything he can to get us out of his way. He may appear to do the job required, but if we look closer, we will see the damage he is doing behind our back. The craftier he is the more covert his actions will be. Initially the decision may seem like a good one. Over time, we might even find our position in the company undermined.

Faker Friends

Fakers have to get close enough to people in order to betray them later. For this reason, it is common for us to end up with one or more of them as a close "friend". He will betray us eventually if the circumstances are right. This may take years or decades, or it may never even happen in their lifetime if no opportunity presents itself for the faker. It may seem that after spotting a faker friend we can at least limit the level of trust, and thereby prevent a big betrayal. This may be true under certain circumstances, but it will be difficult to manage. We would have to make sure that such "friend" never meets our employers, partners, or clients or any other important party in our lives. If they do, eventually they will cause problems – guaranteed. Nevertheless, with acute awareness we can at least minimize the damage. This is a potentially lifelong burden that we all should decide for ourselves, whether we wish to carry or not.

A connection to a faker means many potential pitfalls. They spread their own views, methods, and advice as to how we should deal with life. If we accept any of their ideas, methods, or views we could be "infected". As a result, when connected to fakers, we may pick up some of their negative traits. These traits will not be as frequent and consistent as the faker's. This is why it is important to observe the signs in combination. A true faker will eventually show most signs frequently, while someone who just simply picked up a suggestion or behavior from a faker will only show one or two signs sporadically.

For example, someone may pick up some dating tip from a faker "friend", which comes off as *out of place flattery*. When you look closer, you will find that this person for example only does that when dating. Fakes don't have such selective reasoning attached to their activities. Such copied traits, behaviors, and methods, don't have their basis the same as the faker's rule #1. The copied faker-like actions are a result of accepting a suggestion or mimicking a faker who pretends to be well meaning. The copies don't emanate from rule #1 and thus won't have anywhere near the same frequency of occurrence.

If we spot a faker, we can avoid adapting any of their views, advice, behaviors, or rules. By the time we run damage control, and protect ourselves from the negative influence of the faker, we should ask ourselves what is the point of such friendship? Friendship is supposed to be about mutual trust. Those who still carry on with the "friendship" of a faker for whatever reason can at least reduce the damages by being aware.

Fakers routinely accuse, and mistrust, while they are the ones who, in the end, betray us. They carry out many harmful deeds daily in our societies. Since most of us don't know about them yet, we don't know who to trust. As a result, the victims of a fakers may become mistrusting. Using these signs can help us out of this paranoia, as we can now spot fakers, and know whom it is we should trust.

We can apply the signs in reverse also. To find an honest and caring person, all we have to do is find someone who doesn't display the signs. They are safe to trust. Such a person will make us feel like they possess the virtue of having the exact opposite mentality of fakers. Before we engage in partnership, long-term friendships, or intimate relationships we can use the signs to tell if the person is trustworthy.

While this is not a guide to finding soul mates, it makes a lot of sense that before entering any relationship, step one should be to find out whether or not we can trust the person. Since it is required for the signs to be present frequently and in combination, it takes time to detect fakers. Consecutively finding honest people also takes time. The duration depends on circumstances and the frequency of our observations.

The more we practice and observe the easier it gets.

Blind Spots

Good drivers know about blind spots and observe them at all times to stay safe on the road. Blind spots limit the driver's view. Ignoring these can be especially dangerous when changing lanes or backing up. Not being familiar with blind spots is a sure way to get into trouble on the road.

What makes blind spots especially dangerous is that most vehicles have rear and side view mirrors. This may lead to the assumption that using these mirrors the driver can see all they should. The appearance is misleading. The assumption that you can see all you should see can have dire consequences. Similarly, it can be devastating to fall for the trustworthy appearance of fakers. Warnings alone are not enough. It is necessary to take the warning seriously enough and really grasp the danger. To be certain that a new driver understands the dangers of blind spots it's best to demonstrate just how much can hide in a blind spot. Similarly, to understand how much problems fakers cause it is best to research and understand great betrayals of history as well as recent corporate and government scandals.

Fakers should not be underestimated. We should dedicate the appropriate effort to realize the dangers. Depending on individual experience this may take more or less time but in light of the possible adverse effects, this is time well spent.

Many other conditions exist in various fields of life where not knowing or ignoring something crucial can have serious adverse effects. The common denominator is that a condition exists in each of the situations that promote a false assumption. The required skill set for people varies with the surroundings and activities. There are *must-know* skills and *nice-to-know* skills.

53

Depending on the activities we get into, whether for work or recreation, we should know the specific potential traps, threats, and blind spots. Mistakes are said to be part of learning, but there are certain activities where there is no room to learn from our mistakes. In such case, it is necessary to accept the reality of such threats as presented. Being able to learn from the mistakes, warnings, and knowledge of others, is essential for such activities. Where mistakes can cause great damage or injury, it is necessary or even mandated by laws to observe and steer clear of such dangers at all times. For example, the law does not take reckless driving lightly. It's not like one can get away with a warning or light citation after hitting their "first blind spot motorcyclist". "Yeah ... we understand ... it's your first time ... you'll know better next time ... here is your warning" isn't how it would be treated. (Hopefully in most places)

Similarly since fakers lack consideration for others it is not likely that they can be sweet talked into undoing the damage they have done. What they do can have irreversible adverse effects.

Wherever currency is used, it is necessary for all to be able to keep track of their finances. Without basic math skills, one would be lost. Therefore, basic math is a *must-know* skill for all who use money. Knowing how to rent or buy a home and deal with utilities is also a *must-know* aside from the obvious that speaking the local language is also a *must*. These are several *must-know* skills one should possess in every country. On top of this, one should have some marketable skill, to earn an income. *Must-know* skills are necessary in any modern city for survival. The lack of such skills represents danger and threatens our survival. Even as a tribe member, one must at least know how to hunt, fish, or collect foods that are safe to eat in order to survive.

Learning to play the piano or to get good at tennis would be examples of *nice-to-know* skills for most people except for those of course who play professionally. *Nice-to-know* skills are those that our survival doesn't depend on. We could consider the knowledge about fakers as a *"nice-to-know"* social study. We may even get lucky enough never to run into a faker in our whole life. Such consideration may prove to be as dangerous as ignoring vehicle blind spots. "I don't have to know about fakers" is similar to saying "I don't have to know about blind spots".

The ability to spot fakers should be considered a global *must-know* skill if we wish to make our society fair. When *must-know* skills are ignored, bad things can happen. Indeed bad things are and have been happening globally throughout history.

Human Blind Side

In this chapter, we analyze our blind spot to fakers with the help of an analogy.

All computing devices whether it is a smart phone, laptop, or desktop PC have displays. The display allows us to interact with many applications such as games, email, and word processors. The most important function of the display is to allow the user to see what is going on inside the computer. This is the number one priority of the display. Is the device on? Is it ready to start my app (application)? Is my app still loading? Is my app running? Is there enough memory? Is the battery charged? What's it doing? These are just a few examples of the questions we can answer about the internal condition of the device by looking at the display. We, (humans) also have a display. Is he scared? Is he happy? Is he content? Is he sad? Is he confused? By looking at someone's facial and body expressions, we can answer such questions and more. Our facial and body expression are the display of our mind. The display indicates the current condition of our mental state.

Unfortunately, both of these display methods can be cheated. Computers can be infected by hostile programs (such as virus, spyware, Trojan etc.), which can perform harmful activity without showing any signs on the display. In such case, we could say that the display of the computer "fakes friendliness" as it appears content, while the hostile program may be at large sending our bank information and passwords to the hacker who created the harmful program. Since we don't see the hostile activity on the screen we assume everything is fine.

Similarly, some people can be fakers and plot schemes against others while displaying friendliness. They can smile and put forth

57

friendly body motions, which is a fake display of trustworthiness. Interesting coincidence is that a lot of times a virus ridden computer can be spotted by its delayed response to the user just like the communication delay of fakers mentioned before (in chapter Sign #4: Not answering the question). The extra covert activity seems to take its toll on both humans and computers.

Indeed our expressions can be faked and actors do this all the time, except there is nothing wrong with that, since we all know they are acting. All of us possess the ability to act. Some are better at it than others are. Some may choose to avoid it, but we all have the means to display false emotions on our facial and body expressions and behavior. We can pretend to be happy when we are in fact sad, or vice-verse. Whatever is going on inside the mind isn't necessarily what we display to others. When acting and hostile intentions merge, we have trouble. The person who sells his car with a calm smile, while he knows the engine is about to die, is dishonest and harmful to others, and so is the mortgage broker who does not warn the client of the possible payment shock on their variable APR mortgage. This method of trickery is all too common, and all unsuspecting honest people are subject to fall for it.

This is the human blind side.

Why we Fall For It

Honest people can be unsuspecting for various reasons.

Reason # 1: "*Everyone* is Honest,
Others *DO* Matter"

We know how we think and behave, better than anyone else does. By studying other people, we can approximate their sense of reality (the way they perceive the world), but we tend not to take the time because the importance of doing so is not widely known. Yet, the more we study the reality of others the clearer their motivations become. Since no two people are the same, we can behave differently under the same circumstances. This doesn't mean that there can't be similarities between us. Decent and honest people for instance, have a general concern for the wellbeing of others not just towards their close family members. As a result, they do whatever they can to avoid involvement in shady deals, or imposing anything unfair onto others. This is an important aspect of our reality.

Devious people are not real enough to most of us. Waiting until they betray us isn't the best way to get the reality of them. Studying them and preparing ahead of time makes more sense. If we are unfortunate enough to be the victim, then we tend to have resentment about the person who victimized us. Naturally, the resentment can reduce our willingness to look and understand why people like that do as they do. Our natural tendency is to turn away. This by itself is not enough to avoid similar bad experiences in the future. In order to stay safe, it is more effective to face the problem, and understand why and how people like that operate. The basic mentality of honest and caring people is as simple as, "others *DO* matter." When we live according to this principle, then all other aspects of truth and decency will flow from it. When others do matter, self is *not* elevated above others. Honest people don't feel good about doing things that adversely affect others because others matter also. If others *do* matter, one doesn't trick others into devious or unfair deals. Instead, one will make sure others receive fair treatment, exchange, and respect. When one lives this way, there is no need for hidden agenda. When you deal with a fair person, what you see is what you get. If he or she smiles, you can believe it. If sad, your help just might be required.

Reason #2: Expectation of Fairness

The majority of us are honest. We may fake a smile every now and then, or tell a little (white) lie when forced to, but in general, we seek to avoid that. Even when we do, it's usually with good intentions and in order to prevent upsets. We assume that since we are fair, we deserve fairness from others. Even if we are generally aware that there are some naughty people out there, we don't expect to fall prey to them, because our sincerity is what is real to us, and naughty people are not real enough. Honest people know they care about others, and wouldn't do anything unfair to others. Therefore, they expect the same fairness in exchange, which is exactly what they would deserve. Unfortunately, those who are devious don't care about that. We expect fairness for fairness, which while it is a reasonable expectation, it is irrational in our current society because we do have plenty of fakers. The reality is that we can't automatically expect others to be fair. Well, we can but then we set ourselves up for possible losses.

Reason Number 3: "It Won't Happen to Me"

The news of someone else betrayed does *not* directly affect us since we are not the victim. The event loses some of its reality no matter how detailed the report. In these cases, we tend to assume that it would never happen to us. We might reason that we would have seen it coming, or the person had it coming, or that we just don't know the whole story. Perhaps the victim is not as innocent as reported. We might also think that the chances of something like that happening to us are slim. These thoughts do not necessarily occupy our full attention but linger in the background working in combination with one, or more, of the other reasons we assume. In reality, most of us don't have enough information about fakers, and therefore we don't give it a great deal of thought. Therefore, the possibility of something bad happening to us seems remote. This can cause us to become unsuspecting victims.

Reason #4: There Aren't That Many Dishonest People in Our Society

The percentage of dishonest people is frequently underestimated. An accurate percentage is not easy to estimate, if even possible, as it changes from place to place, and varies with the level of corruption over time. Honest people outnumber dishonest ones, but we tend to judge based on the news we read about the number of criminals caught and scandals exposed. We only find out about a small percentage of dishonest people in our society: the big stories that make the news.

Most of us don't interact with convicted criminals. Since they are locked away, we know little about the damage dishonest people cause. Not all cases make it to the justice system, and even fewer make major news.

For example, in the case of a workplace conflict that results in the unfair dismissal of an employee, the victim might decide it's not worth suing. While to do so would serve the greater good, preventing unfair labor practices, most honest people only resort to the justice system when they have no other option. The expense and time involved in lengthy litigation, as well as possible emotional trauma, are frequent reasons why we prefer to avoid the battle. Yet, for these and other reasons, many unjust deeds go unhandled and unnoticed.

We have different life experiences, and might estimate the honest/dishonest percentage differently. Those who have many bad experiences with dishonest people are likely to estimate that the percentage is high. Others are going to be more optimistic. The exact percentage isn't that important. What *is* important is not to underestimate the number of dishonest people in our society.

Fakers and Downfalls

It is nice to think that every person is genuinely good, and that each of us is decent, and can be trusted all the time. However, this is not the case. Everyone has, or at least had the potential for decency. Some lost this quality, due to a myriad of reasons.

History shows that we have been trying to solve our social and political problems in just about every way possible. There are plenty of gruesome examples of radical government changes, attempts at world domination, decimation of humanity due to religious, racial, or ideological views. None of these attempts worked well and as a result, history demonstrates plenty of downfalls. Even today, throughout the world, racial and religious tensions remain. Racial or religious stereotyping has never solved anything. Those who bring about conflict on this basis claiming a resolution, ignore the fact that these groups are comprised of individuals. Since individuals differ from one another, people within the same group are not all the same. In any large enough group, throughout history, one can find perfectly honest and caring individuals, as well as deceitful liars. Even if we look at one of the most resented group, the Nazis, we still find that a few were just caught in the wrong place at the wrong time. When they realized what they got into, they did what they could to do the right thing. Whether an individual is good or bad (towards others) has nothing to do with their origin, physical appearance, or religion. It has to do with the individual's mentality, hidden from our sight. This is why it is necessary to detect fakers by their actions using the signs. The signs inform us of the hidden mentality.

As far as downfalls go, the current economy is in trouble, globally, and we are facing challenging days ahead. How challenging? Only time will tell for sure. When we look at the current condition of humanity, on the surface it appears we are at peace or at least headed that way. It is only when we take a closer look when we notice a world still filled with simmering conflict – a less than ideal state. We may wonder if our current troubles are due to the same cause as the troubles of the past. If so, is it possible that we've missed something critical? The cause eating away at our society is so simple it's bound to be overlooked: a growing group of unremorseful, compulsive, devious liars whose trickery harms others and spreads like a virus. That's it.

So how can we identify these individuals? They come in all shapes, sizes, and forms so there is no visible way to detect them.

Let's take a closer look. If we considered that anyone who ever lied is a liar, then just about everyone would be a member of this group. Most or all of us lied at least once in life. Such group would not be of much use so we should narrow down the criteria. We're looking at a group of compulsive liars whose aim is to gain trust, and then betray that trust. For example, a good actor, who gets off the stage and lives life without pretending, or tricking others, is not a compulsive liar or faker.

The dictionary definition of faker is:

A person who makes deceitful pretenses.

The extended definition of faker as used in this book is:

A person who makes deceitful pretenses routinely, with no regard for others, or while harboring hostile intentions.

To be more explicit, a faker:

- Does not care about others, but keeps it a secret.

- Displays false emotions to gain trust.

- Exploits trust when the circumstances are right.

- Does all of the above frequently.

Spotting Fakers

The idea of not tolerating liars is not new. In most modern countries, the law regards liars as an anathema. Once caught lying under oath, chances are that person's word will never be trusted in court. (Not to mention other possible repercussions and punishments) The law has zero tolerance for liars. People's possession, freedom, and their life can depend on testimonies. A false testimony under oath is lying.

Thus, it stands to reason that we should treat fakers with zero tolerance in everyday life also. Many bad deeds never make it to court for a number of reasons. Common reasons are that taking an issue to court is costly. Gathering evidence, after the bad deed is done, isn't always possible. What honest person seeks to enter such arena willingly?

One good way to think of fakers is to imagine that their native skill is not what they appear to be doing for a living. Instead, their native skill is the art of faking, which is deceitfulness, scheming, and pretending to be nice and friendly, while having hidden agendas and caring about nothing but self.

It is hard for us to confront this fact. Many of us don't like to look at the reality, since it is not pretty. The reason is that we don't like to be dragged to such level even to look around. In order to deal with them, and understand them, it is necessary to look at their twisted reality, which is not something we like to do. Since we are not aware of the reality of fakers, they take advantage of us. They did that throughout history, and do it now as well. It is unreal for most of us that fakers can pretend to be friends for decades, and then turn on us all in a sudden. What kind of person would do such a thing? – We ask inevitably.

Fakers may appear to be dormant to some but in reality, they are active all the time. Only when we don't know about them, and don't understand that faking is their life style, is when it seems like a sudden change as they betray us. Only later do we realize that a connection to a faker is like an accident waiting to happen. They are busy creating their illusions all the time – we just don't see it until it's too late. Even after a betrayal they will explain how that is not as it seems.

Understanding the mentality of fakers allows us to have reality about them. Only then can we protect our organizations, nations, children, and ourselves from the damages they cause.

We are not looking at some "evil personality", though in some cases the first look at some historical events would suggest that pure evil must be the root cause. Fakers are just trying to survive and move ahead in life, like everyone else. What is missing is their compassion for others. Indeed fakers usually get the idea early on, that by deceit they can get what they want. In fact, they think that deceit is the only (or best) way to get what they want. They think everyone else is a faker and perceive people around them as uncaring. As a result, rule #1 becomes their accepted belief. Their logic tells them that if nobody cares about others, why should they? This serves as the basis of how they think about and act towards others. Because of rule #1, from their viewpoint, deceit as a way of life isn't something to be frowned upon. Ultimately, deceit becomes their most reliable survival skill, and rule #1 becomes their most basic understanding of human nature. These are the underlying thoughts that cause fakers to do what they do, and end up with a synthetic personality. Lack of compassion is an important ingredient of the faker mentality, which is rooted in rule #1. We could get into the many different ways of how one ends up with this view of life, but instead we should only to look at the common denominator. Growing up in an environment lacking compassion is a common denominator. One more thing is necessary though: to be a victim. Growing up in an environment where others are exploited without remorse, is the necessary combination for a negative view of life. Rip-offs and betrayal when repeated can cause the victim to develop the idea that this is the norm. Financial rip-off is far from being the only form of betrayal. Feeling betrayed can begin in childhood with parents, relatives, or friends repeatedly not holding up to their promises. We form our view of life based on our experiences in our attempt to adapt to our surrounding. As a result, when betrayal and deviousness is what a person observes mostly, it can become the accepted view of life. This view is based on a fixed idea such as: others can't be trusted. Views based on fixed ideas are rarely accurate because the observations the idea is based on are usually incomplete.

For example, there are plenty of honest people in the world. Therefore, the idea that no one can be trusted is false. Some people can be trusted while others can't. As a result, the fixed idea that "others can't be trusted" is false. Even in the worst conditions, one can still find trustworthy individuals. All we need is one good example in our lives. A single source of decency can be enough to lead us out of the darkest places. In the unfortunate event when one doesn't have such a beacon of light to guide them, the false view can be formed that "no one can be trusted". A view based on such false observation acts as a filter, through which one views life. That filter will make honest people virtually invisible. If one accepts the idea, that no one can be trusted then what they will see is untrustworthy people even if they are looking at a saint. Over time this leads to a downwards spiral as what they see through the filter validates their accepted view. For example, if a faker is looking at someone making a charitable donation he will assume that the person (or company) has done so because of a hidden agenda. The hidden agenda must be, to get even more back in the long run because of the donation. That is the only way a faker would do such "wasteful" act, therefore, that must be the reason for others as well. Meanwhile an honest person would just look at the donation and think it was out of goodness. One only becomes a faker when enough fakers are surrounding them, and thus the reality of faker's is the most prevalent reality observed. Then the fact that everyone else is a faker becomes accepted as a lifelong view.

This doesn't mean that everyone in such circumstance becomes a faker. Intelligence and education also plays a part in the development of values. There are many documented examples of people overcoming cruel childhood conditions, and still become decent and caring. One good example can be enough for someone to see the light, and find their way to decency, instead of accepting faker's rule #1. The prevalent nature of people is to be logical and be able to spot outpoints. It takes more than just one deal gone sour for someone to become a faker. Most people manage to see the greater picture even in challenging conditions.

For fakers there is no middle road and there is almost zero reality on honesty. As they see it, the world is one big melting pot of nothing but fakers all pretending. In all probability, when fakers read this book, they will see it as another rival faker babbling on about some honesty nonsense, while having some hidden goals. It is important to see that honesty does not exist in their subjective reality. It's a black or white decision. Others either matter, or they don't.

The specifics of when and how one becomes a faker can vary greatly. Most if not all have the bases of their view trace back to early childhood but they only become fakers once they accept rule #1. (This is rarely a conscious decision) Once this fixed idea is accepted other thoughts will be built upon it. It is the nature of the mind to build new thoughts upon existing ones. Knowing this makes it possible to trace back the actions of fakers time after time to the same roots. Fakers may appear to have friends and seemingly working relationships. For a faker, the purpose of any relation is to serve his own desires. It is usual for them to have poor opinions of their partner especially behind their back, which is a reliable sign. (Falls into the category of Sign #3 Bearer of Bad News) Their true nature also shows over time as lack of passion and respect. It's important not to confuse their dormant stage, and a seemingly ok relationship, with honesty or trustworthiness. Regardless of when rule #1 is formed, for fakers it is the earliest and most prevalent rule concerning others. If they had not received so many influences from other fakers, they would not have come to the callous conclusion about others. If ever pondering on the subject they will look back upon their own past as having been dull or naive to think there were any honest people out there. To a faker everyone else is faking and pretending to be happy, content, or joyful, but in reality they are all out to get something. Fakers know that a smile can cover up hostile intentions, so when they see someone else smiling, they believe that beneath the smile they are also after the same things: money and power. When they meet people that they can exploit, they don't view them as honest, instead they will think of them as weaker fakers, who could be easily outsmarted. Since they have no sense of guilt, the more they can get away with, the more superior they feel. To them honest people are just a bunch of other fakers who are easy to trick. What they don't see coming is how quickly their powers will vanish once we become aware of them. The only reason they can thrive is because we've been allowing them, since we are so honest that we don't even like to look at their nature.

Fakers may, or may not be caught in their act. Those who aren't so good at consistently lying are easily discovered, but even then, we tend to dismiss their behavior as a temporary character flaw. We don't realize that what we have observed is the true personality.

When busted, caught in a lie, or a shady scheme, they work on becoming more careful and clever instead of changing their views and ways.

Why that is, is simple. Usually, by that time, they have lived a significant portion of their lives as a faker and deceit is their best-known tool for survival. It governs their reality, which gets stronger and stronger, every time they manage to get ahead in life by devious means. Every deceitful act builds and strengthens the view that this is the most successful tool they ever had and can count on and always fall back on. As time progresses this state becomes more difficult to salvage, as the original decision concerning other people isn't available for them to change. In order to change that, they would have to admit that their most basic accepted tenant, rule #1, is false, and accept the opposite. (Others DO matter) Rule #1 is mostly a sub conscious one below their level of awareness, and therefore it is not readily available. In order to handle it, first they would have to see it. If they were to do that, then every deceitful act they have ever committed would all of a sudden seem like the most treacherous acts ever. In the absence of this epiphany, they don't consider the bad deeds bad at all.

Bad is a relative term. What is bad for one person might be good for another. Someone going out of business, and having to sell out their shop equipment cheap, is another man's "lucky" day for picking those items up below market price.

Fakers just simply lack the ability to consider what is good for all, and are concerned only with self. If we looked at life from their viewpoint, they've never done anything wrong in their lives. When they inflict harm onto others, there is usually someone or something else to blame, and if all else fails, they write it off as, "It's a jungle out there", or "It's a cold, cruel world". It's not easy for fakers to confront their own deeds, as they don't really see anything wrong. If they did look at it, it would overwhelm them because of all the evil they have perpetrated on honest victims. It would be nice to know that there was a way to help them climb out of their condition, but for the time being, there isn't. The only way to deal with them is to control them by withholding power from them. Even if attempts are made to help fakers out of their condition, the faker will probably *fake* getting better. After all, how could they not? Faking is their most reliable tool for it all.

One might think that fakers knowingly and systematically, exploit honest people but it is not a conscious activity for all of them. No two fakers are exactly alike. Their basic rule might be the same but they are still individuals just like the rest of us. As a result, the level of their deviousness and their abilities vary greatly. The more conscious they are the cleverer they can get with their illusions and the higher they will set their targets. Most fakers are fakers for life. The more they get away with the more they will assume that everyone else is just not as clever as they are. What they know is that people can be easily deceived. They don't see that this is because we are honest, they think we are simply gullible. This helps them to develop a superior attitude towards the rest of us. If they just keep up their act, and keep pretending to be nice and trustworthy, eventually, they will gain someone's trust. It's like fishing. Throw in the bait, eventually you'll catch something.

Spotting Fakers

When a faker gains someone's trust, one of two things will happen:

    (1)   They may continue pretending and complying until they are trusted with even more power.

    (2)   They *cash-out*.

(Cash-out is mostly self-explanatory but it will be detailed in the next chapter)

If they continue to pretend, it becomes even more crucial to look for the signs, as they may appear more trustworthy, by not *cashing-out* with the current trust. The clever ones use this trick intentionally. We should pay special attention to how they use, or abuse their power. (Especially when they think, we aren't looking) Similarly, one who is stuck with a faker "friend" may think that the person is a real friend, because of being trustable with small everyday issues. For this reason, it is more important to spot the signs of fakers, instead of falling for temporary indications of trustworthiness. It's one thing to be betrayed by a stranger but when someone we consider as a friend does the same, it can be devastating. It can (and usually does) make us introvert and even consider how and what made us deserve it. As irrational that may be, we tend to do that because we are honest, and think that we must have done something along the line that was unfair. The damage caused by fakers is not only material. In fact, many times the mental anguish they cause with their betrayal is worse than the material damages.

Good is also a relative term. For example, a criminal getting busted is good for the rest of society, but bad for the criminal.

Good and bad are relative terms based on viewpoint.

Initially, it may appear that from the viewpoint of a faker, their actions are not the least bit harmful. After all, this is how they operate, this is their truth, and this is their most useful survival skill. When we look at the big picture, it is obvious that we all rely and affect one another in this world. We share the air, the land, the oceans, the roads and more. We are also subject to the negative effects of a poor economy.

Only a small percentage of fakers "make it big." The majority of them scheme through life, wreaking havoc in their surroundings. Over time, as conditions become worse, there is an increase in faker activity since their unfavorable deeds may cause others to do the same.

Understandably people can and do get charged up on the subject of being taken advantage of. The following analogy may help diffuse some of that charge.

Instead of feeling hate, upset, or revenge, towards fakers, consider that their handicap is that they cannot see the big picture. In that sense, they are similar to children who are too inexperienced to get it. Now imagine being out in the middle of the Atlantic in a large wooden boat, with children running loose with axes chipping away at the hull. They don't get it yet, that the boat will sink if they don't quit. They are just kids wanting to play. You are the adult who gets it, that the boat will eventually sink. Therefore, you should control them if you are to survive. It is imperative to teach them, or if they are still too young to get it then just take those axes away. Consider thinking of fakers as children for life. They just didn't connect the dots like the rest of us on the subject of social living. Similar to the above example in real life, fakers have the ability to sink our society into recessions or wars. We can gain control over them by informing ourselves. If we don't, they will cause us to sink, time-and-time again, as they have done throughout history. The equivalent of "taking the axes away" in real life is to withhold or take away power from them which they cannot handle responsively.

Just as most honest people don't know about fakers due to lack of reality about them, most fakers don't have reality about honest people either. It's not like they're all walking around with a "Ten Signs of Honest People and How to Exploit Them" guide and use that to exploit us. (Thou there are some books out there, which can be used to such purpose) They also fill in their missing reality with the assumption that everyone else is like them. This is why they tend to think that everyone else is a faker. Since they are all individuals, some of them may be in confusion about their own nature. Others may be fully aware of what they are doing. The rest is somewhere in between.

The Cash-Out

Cash-out is a term used to express the final harmful deed of fakers. This is a broad term covering not just financial but all other negative activity. Negative activity means anything that is detrimental to others around the faker. Fakers will *cash-out* when they feel they can get away with something. This can be anything from workplace slander to large-scale corporate fraud. Of course, fakers aren't all the same. Some are manipulative and plan things ahead of time while others just mess things up as they go. The moment of *cash-out* doesn't necessarily mean a fraudulent, or criminal act. It can be any form of betrayal. Before that can happen, they must establish trust.

Having a job offer is already a form of trust. The less obvious betrayal usually happens when a faker is hired. That individual may be delivering *some* of what is required, but at the same time, will tend to slander co-workers, and cause upsets within the company, as he tries to climb higher in the food chain. What he can't deliver, he covers up with lies and illusions. Due to the indirect and sometimes hidden damage they create, he or she tends to cause more problems than solve. For the company this is like taking one-step forward and two steps back. This is still *cash-out* as the faker is continuously getting away with fooling the employer into thinking, that he is fulfilling the required work and moving the company forward. The clever faker, who has higher goals, may wait until reaching a position of great trust before *cashing-out*.

What about Criminals?

Criminals are not in the spotlight for a simple reason. Not all criminals are fakers, but all fakers do commit harmful, unethical, or criminal acts. These can be anything from workplace unfairness to Enron size scandal (*cash-out*). Many never reach any significant *cash-out*, but that doesn't change their tendencies. It simply means that they weren't clever or lucky enough to reach a high enough position of trust to cause a large enough scandal.

Not all criminals are hardened and notorious, and therefore many crimes are unethical or violent attempts to solve a problem. The reason for criminal acts is usually something simple such as financial issues. Many of these crimes are committed in the heat of the moment, when the criminal believes there is no other alternative. Criminals are also subject to the powerful influence of fakers. Many times the person who commits the crime is not the one who thought it up. Since they are so good at hiding their tracks, the legal system doesn't always catch them. Criminals are not the main cause of our problems. Most criminal activity is a byproduct of our faker-infested society. The true underlying cause of trouble is fakers. Through their selfish acts, they infect others and promote unfairness throughout society, which in turn is a perfect thriving ground for crime. Once we handle these issues, worldwide, criminal activity will decline. Just by handling criminals, fakers will never vanish. They haven't in any known society to date. When conditions are improved and people have more freedom to survive in any area, crime tends to decline. The problem is that improved conditions also attract fakers immediately to that area and they slowly but surely decay that improved area over time. It is not a pretty analogy but this is similar to what parasites do. They get attracted to areas of success but because of their faulty views, they end up killing off the host. (The source of success) While it is still necessary to handle criminals, we should inform ourselves about fakers on a global scale so we can take responsibility for them. Nations who rise above others attract fakers very strongly, stronger than it does others.

One faker is enough to ruin a whole government or organization, given that he or she makes it to a high enough position. Direct damage to the organization and consequently, degradation within the organization, will eventually affect the quality of the company's produced goods. Exchanging the inferior goods shortchanges the customers.

Governments are no exception. They also have responsibilities of exchange. Such as maintaining roads, national security, schools, police protection, etc. These services and responsibilities are the government's products and services to the citizens.

When people pay taxes, as usual, and get less and less back from the government because fakers channel out the finances, citizens get the shortchange. At the bottom of all these issues are usually fakers.

Our problem is, and always has been, that we provide breeding grounds for fakers, by being decent and unassuming. There are exceptions of course but most of us don't know enough about them, which is why they can thrive among us.

As they thrive, they lower the overall conditions of society, which in turn causes crime to rise. It's necessary to handle criminals but it is only a symptomatic treatment. We should treat the root cause instead.

Let the System Deal with Them

Some of us want to be police officers while others don't. It's a good thing because if everyone wanted to be in law enforcement, we'd have a real problem.

The fact is that some of us are cut out to be police officers and some aren't. Those of us who aren't may wonder why we should become one now and watch for fakers.

Just as drivers should know how to drive, people living in a society should know the basics about human nature if we truly wish to learn from history. We can't expect the police to be there for us all the time. Since fakers do live among us, they can pop up anywhere and negatively affect our everyday personal and business lives. We can't expect law enforcement to control them all. It's not possible.

Fakers are not a threat until they are trusted with power. Since they can't be trusted with power the responsibility of power falls back on us. This is why we should think twice before we give them power.

The law is here to deal with things that get out of hand and enforce some basic control. Just making more laws won't solve this problem, ever.

"Good people do not need laws to tell them to act responsibly, while bad people will find a way around the laws."

(Plato)

Fakers should be stopped from getting into power not just in the largest of organizations, but in small companies as well. The collective group of all small companies affects billions of lives, as well as the overall economy. It is not just the large scandals like Enron, which we should worry about. We should be ready to withhold power from fakers at the smallest of organizations as well.

It is important to explicitly express that absolutely no violence is intended nor implied. All that's necessary is education, and spotting those who shouldn't be trusted. We don't need people going out "faker hunting" or any nonsense like that. We've had plenty of that type of irrational behavior in history already, and the last thing we need is another witch-hunt. There are usually better and smarter solutions then violence.

After reading this book or hearing some of the contents, a faker may get the idea to turn this into an ideological witch-hunt. It would be a good idea to keep an eye out for that.

Such has happened before many times throughout history. This is an imminent threat worth keeping in mind. They will pretend to be for the cause just to get some power.

Some people might have been badly "burned" by fakers already, and are understandably upset about it. This book is not by any means, the naming of a new public enemy or target. We can take responsibility for fakers simply, *without violence* by removing their means for thriving. How? Informing unsuspecting honest people to know, and become aware of them. It's that simple.

One who attempts a witch-hunt based on this book; or insinuates that this book promotes a witch-hunt, might just be a faker.

(Or uninformed critic who didn't bother to read or purposely ignored this chapter)

Spotting Fakers

We've had plenty of failed attempts in history, which included much violence that always caused more harm in the end than problems solved.

What will happen to fakers in the future? If we know about them and they can't exploit us, they will have no other choice but to produce something worthy of exchange for survival. They won't be able to betray us or reap the rewards from us. No longer will they be able to take the product of others and present it as their own. If their illusions don't work they will have to resort to what the rest of us do – honest exchange. This will not only be therapeutic it may even heal them from their condition naturally as they will realize that they *do* depend on others. Rule #1 will not be validated any longer. As the devious and manipulative aspects of their personality ceases to be the working solution for survival, the honest personality will come out of the background sooner or later at least a little. This doesn't mean we can ease up on them. We should keep an eye on them period.

Allowing fakers to act as trustworthy, by allowing them to go unchecked in our companies, schools, and governments is the sure road to unnecessary failure, pain, and suffering.

The Information Age

The Internet has made a day and night difference in how we access information. Anyone with an Internet connection has access to more information than could be digested in a thousand lifetimes. What we had to go to the library for, twenty years ago is just a fraction of the information that is now available any time of the day. If the answer is not readily available to a question, we can find the appropriate forums where we can ask for professional opinions.

The information flow on the Internet is fast, inexpensive and dynamic. For any given subject of interest multiple sources are available to the degree where it can become challenging just to decide which one to read.

People can say what is on their mind and share their views regardless of being right or wrong. There is nothing wrong with that since everyone is entitled to his or her opinion. One problem the free flow of uncensored information brings about is that it is not always correct. There is no such thing as a central authority that we could use to verify the validity of all the information. As a result, we are on our own when deciding which information is true and which is false.

In this age, more than ever, a heightened awareness is essential to filter out false-information. We should be able to observe things first and store them, and only accept them later as true when proven. Learning how to tame the automatic assumption capabilities of our mind is now more important than ever. This book is no exception. The facts, theories, and conclusions are based on well-researched observations, but it is still up to the reader to decide whether the information is true and useful or not.

False-information: Example 1

This example is well known; as the metric mishap that caused a great deal of losses for NASA.

Many good engineers might have at one point, made such a mistake, and know exactly how seemingly small false-information can cause big problems. The reason for introducing such mistake is strictly to learn from them as much as we can. For those who never heard of this or forgot, here is a quick overview: NASA lost a great deal of investment due to one group using the metric system to design/build something expensive which was supposed to work with something else designed by another group who used English units. False-information, in this example, came from the assumption that the measurement units used were correct. Both teams worked based on this incorrect assumption.

This example shows that the false-information doesn't have to be a whole lot of information being wrong. The deeper (more elemental) the false-information is on a project or subject the more depends on it. The greater number of items depending on it, the greater damage false-information will cause. It also shows the human factor clearly, which is that when we accept something to be true, we never look at it again to question it. We just work with it.

This mistake caused an estimated $125,000,000 in losses. Countless hours and materials all wasted? Only if we don't learn from it.

False-information: Example 2

In this theoretical example, Company "B" acquires Company "A". In company A one guy – let's call him "Faker" – gets a little overzealous seeing all the new management changes as an opportunity and begins badmouthing another employee (named "Sheep") as an attempt to undermine his position. Prior to the takeover Faker was already eyeballing Sheep's position. He began working against him because he felt Sheep made him look bad just by being so good at his profession while Faker ... well ... faked a lot. Now with the new opportunity Faker starts complaining, indirectly. He puts the heat on Sheep every chance he gets by making negative remarks about him mainly to colleagues, but nothing directly to management. He knows better. Over time, the colleagues develop an attitude towards Sheep, as Faker has successfully managed to pit them against him. Faker makes use of every opportunity to belittle the abilities of Sheep. He twists, manipulates, and manufactures facts as needed.

Slowly but surely this reaches management. It comes off as a group opinion by then, because the other employees don't think so highly of Sheep anymore. Management then makes a decision and fails to look at facts. Swayed by false reports about Sheep, they fire him. Humble as Sheep is, he suspects nothing until it's too late. It's not in his nature to suspect treachery because it is something he would never do. In fact, it takes him days after the layoff to figure out what happened.

Sheep was fired based on false reports, manipulation, and the illusions created by Faker. Company B management only realizes their mistake when it is too late, or perhaps never because Faker will continue to manufacture lies to cover up what really happened.

Sheep was not only competent and loyal but also knew about Company A and its projects more than others. If Fakers is good at his trait they will never figure out why the company's sales have gone south. Because Faker will fill the company with enough false-information that the unsuspecting management will never see through.

So where is the mistake here? First Company B management trusted gossip over production facts and history. Secondly, they didn't verify the validity of accusations against Sheep's personality and thought they are making a change that is overdue, which previous management perhaps failed to carry out. Such decisions can be devastating to a company and they happen all too often. In an extreme case, Company A may even fold due to false-information spread by one person climbing the corporate ladder.

False-information: Example 3

Imagine that you were searching for the speed of sound and a website stated it is 1,127 feet per second at 68°F. It also stated that the speed of sound decreases as temperature rises. You try to wrap your mind around *why* that is but this website offers no explanation.

You go to another website to find out. The next website states that the speed of sound *increases* with temperature. What? This is the exact opposite of what the first website stated.

To be certain you didn't misunderstand, you return to the first site. You understood perfectly. The first site simply stated the opposite of the second site. Dilemma. Which one is right?

You look at three more. Now you have information from five different websites. Four out of the five states that the speed of sound *increases* with temperature. Only the first site stated that the speed of sound decreases with temperature. As a result, you can deduce that the first website was incorrect. It is likely that the creator of the site made a mistake. Since you want to be 100% sure, you contact the creator of the first site and clear up the issue. The creator made a mistake and he thanks you for catching it.

Let's see how much trouble this first website could cause for an engineer who wants to design a sonic tape measure. Such devices use sounds to measure distance. The principles are simple: the further away the object the longer it takes the sound to bounce back from it.

Since the speed of sound changes (increases) with temperature, a sonic tape measure may also measure the temperature.

This way it can achieve better accuracy. In a hot summer day at 104°F the sound travels approximately 3 percent faster than on a cool day at 68°F. This means that if the sonic tape measure does not consider temperature, then it will be off by 3 percent on a summer day. It would be off by three inches on a ten-foot distance. This may be fine for estimating work but hardly anything else.

Say our engineer is unfortunate enough to find and use the website with the false-information. He sets out to make an accurate sonic tape measure, which factors in the effects of temperature. If he uses the wrong information (from the incorrect website) then his sonic tape measure will be even less accurate. The device he creates is only as good as the information it is based on.

The unfortunate engineer is likely to go around in circles trying to find what he did wrong. Everything checks out however. The information was accepted as-is. It came from a source he trusted without a doubt.

Eventually – usually by accident – the engineer will stumble upon the source of the trouble being the wrong formula, which gave him false-information to work with. The device will only get fixed when the false-information is discovered and corrected. Such problems occur more often than we would like to think, causing engineers to waste valuable time and effort on a daily basis. Good debugging skills are essential for engineers, and that is how we discover the underlying cause of such problems.

The knowledge of spotting false-information is not widespread even among technical people. Some seasoned engineers develop a skill for false-information spotting on their own but things would be so much simpler if we knew about such phenomenon right from the beginning as opposed to figuring it out the hard way.

False-information Sources

While false-information can come in virtually limitless forms, it can all be filed into one of four categories based on the source. The categories are:

- False-information due to error.
- Intentionally falsified information.
- Omitted information.
- Missing information.

These categories help realize the source of the false-information, which makes it easier to recognize them in the future. When the source is isolated, the road to correcting the problem becomes clear. Like most new viewpoints, looking at problems in terms of false-information may take some time to adapt. It might be necessary to think with it for a while and see how it applies in reality and to our experiences. Over time, we can use this view and spot the sources of our troubles and problems much quicker.

"False-information" is a general term, which includes all four categories.

The exact definition of false-information for this book is:

Any information, which deviates from reality. It includes missing and omitted information.

Missing and omitted information in this sense means "information with maximum (or infinite) deviation from reality". False assumptions and omitted, missing, altered, incorrect, inaccurate, partially correct, intentionally falsified, unintentionally falsified, damaged, and mistaken information are all covered under the term "false-information". The reason for that is the way our mind handles missing and omitted information. It "makes stuff up" to fill in the gaps. Conventionally missing and omitted information isn't considered false-information. The reason we can afford to consider these false, is that we are exploring the subject from the viewpoint of our minds. Based on what we tend to do with missing and omitted information it is feasible to treat them as false-information.

False-information Due to Error

This type of false-information is due to an honest error.

Examples:

- Typographical errors
- Damaged document
- Misunderstood word or sentence
- Erroneous translation of word or sentence
- Report based on incomplete observations
- Report based on false-information

One thing in common between all of the above is they are all forms of honest mistakes, which passed or bypassed the necessary verification. This means they do not compare to reality well enough.

The first two (typo/damaged document) are self-explanatory. They can happen to anyone. Knowing about these and reducing the occurrence of these is important. The moment we let our guard down one can slip in. For typo's it's important to always observe that what is on the screen is what we intended. This may be obvious but when one types a lot, it is easy to forget to check especially in the presence of external or internal distractions such as upsets, noise, etc. In case of software engineers, a single character typo can render a program nonoperational. Small mistakes can have large-scale adverse effects in many fields.

To avoid errors due to a damaged document, ample protection and backups are required. Many useful forums and tutorials are available on the Internet to help with these, making it unnecessary to cover the subject in details. For the purposes of this topic, it's enough to be aware that information can be lost, damaged, or changed for many reasons. We lose important data due to infrequent backups, theft, lost devices, etc. Estimates project that such losses amount to billions of dollars each year in the US alone. Any data not backed up is an accident waiting to happen. Some cause financial losses, while others cause upsets as in case of lost photos and videos. Data loss will force us to fill in the missing information from memory or recompilation. This is a perfect opportunity for false-information to enter. The pressure and mental anguish the loss may cause can lead to mistakes.

A misunderstood word or sentence is common. People can be upset over misunderstandings even when they speak the same language. This can occur in several ways. The most obvious is that in many languages, words have multiple meanings. When someone says something, the one listening may not be thinking of the same meaning for the words they hear. (The same applies to written communication of course). Another is sentence structure. The meaning of many sentences can be ambiguous. This means we can interpret it in more than one way. This is due to the way we put words together. Example of an ambiguous sentence:

"The girl hit the boy with a spoon." Did the girl hit the boy using a spoon, or did the girl hit the boy, who had a spoon? It's up to the listener how they interpret it. When sentences are misunderstood, it allows false-information to enter our thoughts. Another possibility is that the listener may have a different personal reality (experience) related to the discussed subject. For example, the person who experienced many near accidents may see a news report about an accident differently. They may think: "Another crazy driver, now I don't even feel like going anywhere." While the person who had a problem-free driving experience and considers the roads safe, will see the accident for what it is, one unfortunate event.

Sometimes we simply don't know the correct meaning of the word. This is a touchy area because we don't like to be caught not knowing our native language. To ease any tension, pointing out the miraculous process of how we learn our native language, is useful.

Spotting Fakers

Most of us think nothing special of this, but it is very impressive that we can do this in the first place. Consider it this way: first, we are born. Then we see some giant beings carry us around putting clothes, diapers, and such on us and then removing them periodically. They stuff food in our mouth, make all sorts of funny faces, and flood us with sounds, day in and day out. Sometimes they speak in funny distorted voices. They are also pointing at things while making sounds. Then they point again at the same thing while making that same sound. Aha! A pattern! Our mind starts piecing this huge puzzle together. Despite the funny voices, faces, emotions, we sort it all out. Word-object, word-object, word-object sprinkled with some senseless stuff here and there but we succeeded anyway. Then there is word-action, word-action, word-action, and we piece that together too. We continue observing until one beautiful day we say our first word, and make our parents overflow with joy. Since you can read and communicate, you can be sure that you possess this miraculous ability. How we do it exactly is still not fully understood, but that doesn't seem to stop any of us from doing it, now does it?

While solving this complex puzzle, and mastering some words and later conversation, most words are perfectly understood, but it's only natural for a few to get messed up due to silly habits and the amount of false-information we have floating around in our collective consciousness.

A few words with a foggy meaning, and another few that we just flat out know wrong, doesn't take one bit away from the miracle we accomplished. We've learned a language! It's amazing how little we miss and that we can sort this mess out in the first place. Obviously, we are "wired" very well for learning. The older we get, the deeper our understanding. Our native language is no exception.

Even with the best of intentions from our parents and teachers, false-information can and will slip in, causing us to miss out the exact meaning of a few words. Finding any words that we may be unsure of and looking up the exact definition is a virtue. Language after all is a *must-know* skill. Unfortunately, under certain circumstances we may be influenced to feel ashamed when we don't know the meaning of a word.

Hearing about fakers might seem a bit tiresome by now especially if you don't have a negative experience with them. The word "faker" appears in this book over three hundred times but for good reason. They affect all of us even if indirectly. Seeing many examples of them in different scenarios makes it easier to spot them. This happens to be another good spot to inject something about them.

Fakers can, and do make us feel bad about not knowing something, whenever they can. Even if ever so gently. The truth is they are the ones with a lesser understanding of reality. Whenever they spot that someone else doesn't know something, it is a good opportunity to use that as leverage. Even if there is no material gain in sight, at least it helps them feel and look superior. (So they think)

The ability to learn our first language is so obviously natural, that we don't even notice it. We don't even feel the need to congratulate one another for this great achievement. In reality, we all deserve a pat on the back when we achieve a good command of our native language. If you didn't get one yet, it's way overdue. Please consider yourself belatedly congratulated. Seriously. Just because we can all do this, doesn't mean it isn't a monumental achieving. It is! When we truly grasp the size of this achievement then we can realize how little significance it has that we have to find some false-information, some of which happens to do with the meaning of words. Even if we have a hundred misconceptions, it's all due to false-information. These are nothing compared to how much concepts, words, and skills we did get right. Realizing this can help maximize our trust in our mind and ourselves.

Translation is another potential source of false-information. The expression "Lost in translation" should just about cover it. All of the possibilities that apply to misunderstood words and sentences apply here also. On top of that, we have issues arising from customary, grammar and cultural differences. Expressions and jokes rarely translate well from one language to another. Even within the same language, translation may be required. Explaining medical or technical findings to non-technical listeners is a form of translation also. Explaining a mathematics theory in plain English is another good example. Translation or interpretation errors are a constant potential source of false-information.

A report based on incomplete observation is also frequent. The skills and knowledge of those who conduct scientific research vary greatly. (Due to false-information) Skill, mindset, and attention to details all play a part in how accurately we carry out an experiment, and how well we observe and translate the findings. It's common for some people to jump to conclusion due to cutting corners in their research. Insufficient research will cause us to assume because of the lack of information. Some call this an "educated guess". Whatever the name is, in order to have certainty we should strive to get every necessary piece of information. Then verify each, before drawing conclusions. Otherwise, the risk of presenting false reports and conclusions is high.

A report based on false-information is important because it shows the infectious nature of false-information. For example if someone is preparing a report on a given subject and uses false-information from a source, then every part of that report that relies on that false-information will be false also. Why? Because when someone uses any false-information to build on, that information becomes the foundation of the building process. The greater the report is which we build upon false-information, the greater the fall. If we conduct a research based on false-information, it is like having a house that is shaky, because it was built on a weak foundation (or quick sand), then trying to continue by adding a second and third floor. It will eventually crumble.

Intentionally Falsified Information

Whether it's just slight alteration of facts or complete manufactured lies we all have the freedom to say or write whatever we please. The reasons for falsifying information can be as simple as greed or fear. It can also be the result of complex emotions. It can also be a harmless prank.

Every prank requires false-information. Trivial example: A child might say to another "Your shoelaces are out." The other looks down and while being smacked on the nose/chin realizes "It's not!" To pull such prank two fundamental elements are required: a straight face and a lie. The lie is based on the false-information "shoelaces are out". The straight face (or better yet a convincing expression) is the misleading decoy. Every trick can be broken down to such elemental components no matter how complex they are.

Some tricks involve presenting trusted information, smiles, directing attention, all to cover up one or more lies in the end. Pranks, tricks, and frauds all have these ingredients. No matter how clever or time consuming the prank is, it contains and structurally relies on false-information presented as true.

This is a good spot to reiterate that the phrase "false-information" is like a bucket. Any information with notable deviation from objective reality is 'thrown' into this bucket. Since missing and omitted information have maximum deviation from reality they are also included in the term "false-information". (Wherever applicable)

Deception also draws its power from false-information. Presenting false-information, with the intent to derail the attention of the subject, is how it's done. Sometimes the "prank" is not harmless and the prankster is looking to gain something. In such case, the prankster is looking to hide the prank as long as possible. If the "prank" qualifies as illegal, the "prankster" wants to hide the facts indefinitely. He doesn't want to be caught.

The categories of intentional false-information are:
- Falsifying information with *no* harm intended
- Falsifying information *with* intention to exploit or harm

Such categorization allows us to separate the good from the bad. Not all pranks are bad, or unwanted, and therefore we should differentiate. If we declared all pranks as bad, we couldn't even enjoy a magic show. Even if the intention isn't harm, sometimes things go wrong and there can be adverse results ranging from hurt feelings to a street fights. If things do get out of hand, it is usually because the intention behind the prank is misinterpreted or the intention was to harm. There are also gray areas where the intention was to harm only a little. The level of harm depends on the degree of hurt feelings and losses. The underlying culprit is the false-information. A onetime incident clears up easily when people talk to each other about it. Finding the misunderstanding can restore the original "no harm intended" state. Falsifying information *with* the intention to harm (or indifference about the wellbeing of others) causes much trouble in the world. It is a fundamental tool of fakers.

Omitted Information

The next form of false-information is omitted information which is what happens when people withhold anything from us for any reason. Only intentionally omitted information is addressed in this category. Unintentionally omitted information is considered missing information. When we come across omitted information, our tendency is to assume. (Covered in chapter, "How We Learn")

The less we assume the more certain we get. As in the case of intentionally falsified information, we can also categorize intentionally omitted information as follows:

- Omitted information with *no* intention to exploit or harm

- Omitted information *with* intention to exploit or harm

It would be nice to live in a society, where lying and omission are not necessary for survival. Omission of information knowingly, is a form of lying, which honest people refuse to do whenever possible. In our present society, omission of information is necessary. For example, restricting military information to ensure national security is essential. It is also necessary for companies to protect their technology and trade secrets. On a personal level, we are also required to omit information. For our own good, we are required to keep our passwords, financial and legal information from others. These are preventative omissions. If we don't protect such information we may be attacked, cheated, robbed or defrauded. In our world, we can't just leave our doors open with money laying on the table, and assume it will be there when we come back. In reality, we need to protect our personal properties and assets. Omission plays an important role in this process, and it's something we can't do away with overnight.

The competition among companies requires trade secrets kept well protected. Manufacturing and process methods have to be a secret or protected by patent laws. This affects the field of science all the way back to research. In many cases, researchers omit the full details of a discovery. Research requires financial backing. Usually, the funding for the research is an investment. Sharing too many details would allow the competition to get a head start, without spending the money on the research itself. To prevent such unfairness, companies routinely omit information. People in the fields of technology and science have to be careful with their information. This has an unfortunate crippling effect in the field of science, because the flow of information is restricted, and sometimes intentionally distorted. Since the details of a research are not disclosed, we are supposed to accept the results as-is. Previous chapters already hinted that this is not a good idea. We should verify first, but we can't do that if the details of the research are omitted. When we don't verify something, and instead accept or believe that it is true, we open the door to errors. The widely accepted secrecy in the scientific arena is a safe haven for intentional and unintentional false-information. Organizations may fund research with a certain interest in mind. As a frequent result, proving what benefits the funders receives too much focus. This also introduces the possibility for false-information, in the form of overlooked or biased research. Some of these look very convincing but are flat out false reports. Keeping an eye out for such false reports is highly recommended. It is not a good idea to believe anything just because a scientist in a white lab coat reports it. (No offense, pun, or stereotyping intended) Unfortunately, scientists (as a group) are no exception from the presence and influence of fakers. Fakers can pretend to be a trustworthy scientist. They will knowingly put that white coat on because it is a great tool for their illusions. They can make it in the field because true, decent, and honest scientists don't like confrontation. This allows fakers to positions, they don't belong in. Just because someone is an expert or a scientist doesn't mean they are any less susceptible to fakers. (Except for scientists in the fields that study human nature perhaps)

Spotting Fakers

If we can't verify something because the details are not shared, it is not wise to accept the result as a fact. If we accept it, our mind will be using it in the future whether it is true or not. A faker scientist will gladly take advantage of us, if we aren't trained enough in that science to verify the results. Omission is a frequent tool of those who manipulate others. Regardless of the motive, the troubles omitted information causes, range from insignificant to life altering.

There is nothing wrong with manipulation, if it's done with caution, and making sure that others do not suffer as a result. We influence and manipulate each other all the time anyway. Example: "Want to grab a burger for lunch?" If the questioned agrees, his path alters (is influenced) immediately. Whatever he was going to do for lunch prior to the question changes the moment he accepts the invitation. Maybe he was planning to have pizza for lunch, or skip lunch entirely because he is on a diet. Who knows? Why did he accept? Perhaps just to enjoy the company of the other person. Why was he invited in the first place? Perhaps it is a single female colleague that invited him hoping it might turn into something more. If she does not disclose her true reason for the invitation, then she is omitting information. Is he going to mind? Not likely if he accepted the invitation. The omission may be subconscious or fully conscious. It doesn't matter. No one is about to be harmed in this example therefore the omission is harmless. It is well within the range of what we don't mind to accept. Intended harm (financial, mental, or physical) or the indifference to the wellbeing of the others, is what makes omissions harmful.

Intentionally not sharing information for personal gain, or to harm another, is the concerning subject. Selling a boat with a weakened hull, without warning the buyer about the problem, is one example of omitted information for personal gain. Furthermore, it would be reckless or intentional willingness to harm, if the seller knows that the buyer is planning a deep-sea trip right away. Omitted information is all around us, and causes a great deal of confusion. Being able to spot these is just as important as spotting intentionally falsified information.

Classifying omitted information based on intention is important because not all omitted information is bad. Omitting information with no harm intended is a crucial element of games. We all love to play games. Whether it is a card game, board game, video game, car racing, tennis, football, or the game of life itself, we are all players. Without withholding information, none of these games would be the same or even possible. If we knew the hand of the other player in cards or knew the plan of the tennis opponent, or knew what the other party is thinking, games would not be possible. Therefore, it is important to see that not all forms of omitted information are bad.

Life itself requires some information omitted. We should have our own personal space, property, and life. Doing away with that is not what is proposed. Only deviously omitted information, with the intention to harm, or no regards to others faith, should be curbed. We can do that simply by spotting the presence of these and steering clear of those who don't play nice.

Missing Information

Missing information is when we need some information but we don't have it. Grouping it with false-information is due to our tendency to assume in the absence of information. It has the same potential effect on us as omitted information.

Missing information is a frequent cause of errors. It requires precise research to fill in the missing gaps. If the research is incomplete, or affected by any other form of false-information then we have a tendency to fill in the gaps with incorrect or made up information.

When we meet someone for the first time, we have a great deal of missing information about the person. Their entire life experience is missing information to us. As we get to know the person, our mind begins to fill in the missing information. Since we have the tendency to base our view of others on ourselves, we end up filling in some of the gaps with assumptions. This is why we tend to assume that fakers are trustworthy. We tend to think, "We are honest hence they are also". As they dramatize their trustworthy image, they make our assumption stick.

A good way to deal with missing information is to develop a habit of spotting it and acknowledge the fact that it is missing. Once we do that, the missing information can't give us trouble because we know it is missing. If we need to find out about it, we'll know where to look. Anytime we meet someone new whom we don't know anything about, it is a good practice to make a mental note of this. That way we can reduce our minds tendency to assume. One quick thought such as "This is a person I don't know anything about" can take care of that.

In any situation, it is best to take in things as they are. As our mind tries to predict events for us, it is very eager to fill in the missing gaps. This is useful but can also get us into trouble from time to time.

In any situation, it is best to read only the absolute minimum into whatever we are experiencing or observing.

Example: imagine that you are appraising an expensive vase in someone's home for insurance purposes. Say this vase is in cabinet that prevents you from seeing the back of the vase. It looks nice and intact. Your mind may fill in what the back of the vase must look like. The shape and artwork are simple and predictable. No reason to think the back would be any different. Our mind is filling in the missing information for us. We can practically see the back of the vase with our "mind's eye". Except what if the back is cracked or has a gaping hole? It could affect the value greatly.

In this case "reading the absolute minimum" into the situation would be to think this:

"I see half a vase". (Or 60% of it. Whatever the case may be)

To be certain we would have to take it out of the cabinet and look at the back. If we can't do that for some reason we shouldn't automatically assume that it is a complete vase.

As we experience life, our mind does this automatic prediction gap filling process for us routinely. Whatever we can't see directly our mind tends to fill it in for us. This is assumption! Once we spot this process then we can start taking control of this automatic assumption habit. It is not a physical property of the mind to do this. It is a habit, which we can change knowingly with practice. The less we assume the more we are able to see the world for what it really is.

How We Learn

Most of the skills described so far are usable instantly. Just like magicians, fakers rely on illusions. What's beneath illusions? False or omitted information is always necessary to create and illusion. Why are we susceptible to fakers and false-information? In order to explain this, the remaining chapters will reveal certain aspects of our minds. Naturally, some of the following chapters will be a bit more technical as they fill in missing information about the workings of the mind. Understanding these may take a bit more dedication but the results are worth every bit of effort.

Regardless of the activity or profession we become involved in, when we first start, we are beginners at it. The only exception is when we have already existing experience to build on from another profession we have already mastered. For instance, a finish carpenter can build on his existing knowledge, and switch over to construction framing easily. After learning the applicable laws and codes, he would require little practice because he already has the mechanical skills. In comparison a chef, no matter how good he is at his profession, would have a longer learning path for construction framing. In essence, it all boils down to ample subjective reality (experience). If the activity involves risk, then knowledge, practice, and caution are essential. Constant awareness and attention to every detail is very important. As we spend more time, we get better and better at what we do. Learning a skill occurs by gathering and processing information first to achieve some level of understanding. In order to get good at any skill, after we process the new information, it's time to put it to practice.

Learning begins with communication: discussions, reading a book, or watching a documentary, are all forms of communication.

A discussion is called a two-way communication because people pass information and ideas back and forth. During two-way communication, we can see that the person we are talking to is receiving our message, and we can immediately receive his or her response. Reading a book, watching a film, browsing the net, and listening to the radio are all forms of one-way communication. One-way communication is the term to describe these because information flows in one direction only. For example, reading is a one-way communication because the reader (the receiver of the communication) doesn't have an immediate way to communicate back to the writer (the sender). In this sense, just simply observing the world around us is also a form of one-way communication. Whether we are marveling at the sunset or watching grass grow, we are communicating with the world around us because we are receiving information about it. Just simply staring at a wall is still a form of communication. Our eyes are continuously receiving information from the walls surface (light and colors), and we know "it's there".

During learning, we communicate with the subjects, things and persons involved in that subject. Teachers, books, audio, and video are all sources we communicate with during this process. Considering this definition, it is evident that learning starts with and continuously depends on communication.

After we gather information, we have to process it. We do that by playing with it in our minds. Twist it, look at it from different viewpoints, and use our imagination. The information we gather then becomes part of our knowledge. We can combine it with other information we have gathered already in virtually infinite combinations in order to make sense of it.

It is natural to be unaware of the vast amount to information our mind processes each day. Whether we are learning something complex, watching a show, or just walking in a park, our mind processes all that we see, hear, smell, and feel around us. For the mind, there is no such thing as downtime except when we sleep. Whether we are waiting at the dentist bored out of our minds, and staring at the wall, or studying calculus, our mind receives the same amount of continuous information flow we perceive. Much of the information we gather gets stored in our mind somewhere, somehow. This is similar to copying. We can recall later what we have seen and experienced. What we recall is just an image similar to a photocopy. This is merely a copy of the real thing, the real event, and place, which was around us at the time of copying (recording). As we live our lives, we continuously make copies of the real world around us. These make up our personal mental image collection of the real world. We use these copies in order to think about reality.

Our body has several different organs that we use to sense reality. These organs are sensors to sense the real universe around us. In order to analyze and understand the universe our mind makes copies of it by using the body's sensory organs. The copies therefore contain pictures, sounds, smells, taste, balance, motion, and various physical sensations. These copies will be referred to as *mental-copy* or *mental-copies*. Thus, a *mental-copy* is what we store in our mind about the real world around us, as we have experienced (or sensed) it. Each *mental-copy* can contain picture, sound, smell, taste, and physical sensations all together. The sensory organs will be referred to as *body-sensors*. Thus *body-sensors* includes eyes, ears, nose, taste buds, balance and motion sensing organs in the ear. *Body-sensors* also include the pressure, chemical, and heat type sensors embedded into our skin as well as inside the body. Many conditions are detected from the inside of the body such hunger, being full, content, pain etc.

Body-sensor is a group name for all the sensory organs that the body has, even those not specifically listed.

Our bodies are equipped with a large range of sensors, and we deal with reality indirectly, via these sensors.

Looking at our body in such way can be new and strange, but it's important since it is the first step in understanding our thought process. It's also the key to understanding how and why we can be tricked by illusions from time to time. The world around us feels so real that we seldom think about the fact that we (our minds) are actually working with copies.

All day long, we make a tremendous amount of *mental-copies*, as our mind analyzes all the signals from our *body-sensor*. How much and how it is stored, is a large subject. It appears that while the mind processes most or all of the signals from our *body-sensors* it may not store or even process everything. (Discussed later) The sensors in the skin are mostly pressure, and heat based. Our mind and nervous system translates these into many different sensations such as pleasure, pain, heat, cold, or contentment. For example, too much pressure signals from the same area or too intense heat detected means pain. Our normal reaction to that is to stop it from continuing. If someone gently massages our back, it means pleasure or contentment but, if the person pushes his or her elbow too hard, the signals from the same sensors translate to pain. Naturally, we want the person to ease up.

What is happening around us, or to us, at any given time is reality. It is the real thing, the real universe, which surrounds us. This is not what we use for our thinking process; instead, our mind has to rely on our *mental-copies*, and *body-sensors*. This copy collection is the foundation of our subjective personal. When the mind decides to store the signals sent by our *body-sensor* then all the contents are subject to be stored at will.

The mind is also capable of recording sequences. These sequences are similar to a motion picture, except motion pictures only have picture and sound. The recordings of the mind contain pictures and sounds also, but smell, taste, balance, motion and all other senses can all be stored along with the picture and sound. The recordings, should not to be thought of as one long recording of one's entire life. They are fragments of sequences that the mind compartmentalizes and categorizes according to how it sees fit. Other than what is available from *body-sensors,* the mind records one more parameter as part of the *mental-copies*. That is the state of our mind itself at the time of recording. Think of this as an overall feedback, as to how we were feeling at the time of experience. This can be any emotion: ecstatic, miserable, happy, sad, content, angry, furious, etc. If we like or dislike what we experience, we record that feeling along with the picture, sound, taste, smell, etc. As we live our lives, the *mental-copy* collection keeps growing and growing.

No two people have the same *mental-copy* collection, since no two people have the same exact experience every minute of their life. It may seem like at first look that close siblings or twins would have the same experiences especially if they spend most or all of their time together. In reality, there is no such thing as "exactly the same experience". Even the tiniest differences add up over time and can cause great differences in subjective reality. Since the subjective reality is based on the *mental-copy* collection, no two people will have the exact same *mental-copy* collection either. It may be similar but never the same.

It seems logical that the things that we have many *mental-copies* of, in our mental library, would be the things that we know more about. More *mental-copies* increase our subjective reality. The things that we have more *mental-copies* of, we know in a way that is closest to the real thing, the real universe around us that we have experienced. More *mental-copies* on a given subject or item means more subjective reality, which is the key to deeper understanding.

For example, when we have somewhat mastered driving, we could say that we have a good subjective reality with driving. Some would describe that as good understanding of driving but consider this: For most people driving can develop into quite an automatic activity over time. This makes it obvious that the key is the amount of time spent with the activity. More time means more *mental-copies*, but as we'll see this is not the only important factor. As we practice driving, it's not like we develop a deeper understanding of the pistons, physics, or friction and such. We just drive more and as we do, our subjective reality database keeps on building. This is why it is important to look at reality as a factor and not just theoretical understanding. Touching and operating the shift knob/handle and pressing the pedals become real to us only when we start doing it, which is the part that no amount of reading can replace. To give an extreme example we could have enough mechanical knowledge to build a bicycle and know all about how it works, but until we get on one to ride it, we will have no reality with riding a bicycle. As we progress towards becoming experienced drivers, our reality with the vehicle and the various possible situations also become more real. Danger becomes even more real after a situation, such as having to dodge a reckless driver. The more we drive the easier it gets to the point that one day we don't even think about it. We just hop in and turn the key and the next thing we know is that we are at our destination. The reality one has about driving is the individual's own reality. This type of reality is referred to as someone's or one's reality or in general as subjective reality. This is always somewhat different from objective reality. That's because it is based on copies of the objective reality (the real world) and copies by definition are never the same as the original. The expression "subjective reality" is in use instead of the word "experience" purposely. Thinking in terms of subjective reality helps realize the key to experience, which is gathering enough *mental-copies* of the real world. It promotes viewing in terms of what is real and what isn't.

Here is one more way to look at subjective reality: say you are on vacation and come across a beautiful tree that you are compelled to make a lasting memory of, so you pull out your camera and snap twenty shots. Ten years later, when you are looking at the photos of the tree, you notice something you wish you could make out but you can't. "How could I have missed that? I wish I had a close up of that angle." You are missing an important angle. Having a good reality of a subject would be equivalent to having shot 5,000 (or more) images of that one tree from different angles and heights. Our minds make tremendous amount of *mental-copies*. If we could only record one copy per second it would yield to 3,600 *mental-copies* per hour or 57,600 copies per day. This yields a potential 21,024,000 *mental-copies* per year. (This calculation is based on 8hr sleep per day) Our minds can record more than one *mental-copy per second.* This is not to say that we do record that much. It just means that we have the potential to do so.

When we are beginners at any skill, our subjective reality is very different from the objective reality. As we progress with any learning, the subjective reality increasingly approximates the real world. The more *mental-copies* we have, the closer our mind can get us to reality. As a result, the subject gets more real to us.

Everyone has his or her own version of reality. The driver, who never had to dodge someone else, will think of driving as something safe. This will be part of his subjective reality related to driving. Chances are that his reality is going to be that way even if he heard about some accidents before. We have this habit of "seeing is believing" and it is not an easy one to bypass. It is also telling of the difference between understanding and experience. If the previously mentioned person has to dodge some reckless driver on the road one day, his reality on driving is going to change to something like: driving is mostly safe, but occasionally people do some dangerous stuff, and thus I must be more cautious. Then if this person ends up having to avoid two more accidents within a week, he will likely have a subjective reality such as, "People are crazy, and reckless". "I must be super-vigilant". In the extreme, this person might even limit driving, to the most minimum necessary.

Subjective reality varies greatly from one person to another since our experiences cannot be the same. For this reason, it is useful to keep in mind, that our experiences are not the same as others. If we experience something more frequently than the average, then we tend to see things differently from what they really are. Reading about something, or hearing about it, is not the same as seeing it happen, which is why the saying "seeing is believing" is so spot on.

We can read and talk about bike riding all we want. Until we get on one, it doesn't matter how much we read or talked about it. It is well known that for physical aka motor-skills, theory alone is not enough. There has to be physical practice after a certain point, and there is no way around that. The physical activity may be broken down into its components and one can prepare much better that way. Physical practice is still required to build subjective reality with the activity. For example, balance boards, which aid the development of balancing skills, speed up the ability of children to learn to ride the bicycle. They still need to get on the bike eventually, but learning is faster this way. (Can be safer too) This makes sense, because one of the most important skills for bicycle riding is balance. We can't develop balance by reading or talking about it. Our mind has to receive ample amount of actual *body-signals* from our balance sensors. (These are located inside the body near the ears) Then our mind can build the necessary subjective reality for bicycle riding.

Practice is similar with other skills also not just physical activities. Mental activities such as adding numbers in our heads or mentally visualizing things also takes practice. It takes playing with numbers in the mind, visualizing and putting some hours into it daily to get good at it. Thus for any skill, mental or physical, it is no news that "practice makes perfect". If the activity is restricted for some reason (like flying an airplane), it's a good idea to break it up into its components. Practicing the individual components can get us close to the real thing.

Technology now allows us to take advantage of simulators. Anyone can drive a simulated vehicle at the nearest arcade and get a feel that is fairly close to the real thing. Vision, sound, vibrations, and even vehicle body motion is emulated on some arcade simulators. However, one thing is missing. In order for these simulators to have value as preparation for real driving we should make believe it is real! Arcade simulators could have a realistic practice mode.

Some of us make believe frequently while others don't. Make believe is a useful ability. In fact, it is the cornerstone of creativity. Make believe is harmless, as long as we know what we are doing. That is, we know we are playing make believe. This allows us to know at all times what is real, and what is not. As long as we have the awareness that make-believe stuff is not real, we remain in alignment with the real world and in good control over reality.

Now back to the subject of reality. Reality is simply what is. The physical universe with all its matter and changes is our observable reality. All the other people around us, is also part of reality as well as our own body and mind. Reality is what we can observe, and interacted with.

Subjective reality is what we have observed and assimilated from the real world, and is ready for us to recall so that we can work with it in our minds, to do whatever it is we wish to do with it. Each of us deals with the real world based on our copies of the real world. Subjective reality is different for everyone.

Objective reality is what it is regardless of who is observing. Life experience takes us all on different paths, and we experience even the same object differently. We simply can't be in the same place and same time as someone else. Even when two people are looking at the same vase in the same room, standing side by side, they will see that vase from different angles. The one on the right will see a little more of the right side. The one on the left will see a little more of the left side of the vase. The shadows observed will be different. The room lighting will be different. The one on the right might even recall the viewing later as unpleasant, due to a ceiling light behind the vase, which was blinding him. Meanwhile the one on the left was getting a nice pleasant look absorbing all the details of the artwork. While such differences may be miniscule for just one event, these little intricate differences add up eventually, and cause us to have entirely different subjective realities.

Two people watching the same concert will not have the same reality either. Someone up front will hear the music louder and might even shake hands with the performer, while some other fellow in the back is using binoculars to get a good look and complain that the music isn't loud enough. It's easy to get certainty on just how different the subjective reality of individuals really is by striking up conversations with a few people leaving the same event. Any event will do such as a concert, cruise, or school meeting. To do that, would be useful for anyone wishing to have a good understanding (and subjective reality) with this subject. Being able to see the differences between events (and things) is just as important as finding similarities.

Without sufficient subjective reality on any given subject, we would be as lost as any beginner would be. For example, let's look at someone starting a first job. Just about everything is new such as rules, habits, requirements, and the activities of work. It would probably be easier to list what isn't new. Becoming comfortable will only happen after some days, or weeks, or even months, depending on the individual's adaptation skills, interest and various other circumstances. As time passes at the new place, it will become more and more a part of this person's own reality. As the subjective reality grows, so does the comfort and abilities of the individual.

Subjective reality is therefore the foundation of our knowledge. Building our subjective reality requires that we spend time with whatever it is we are learning. Spending more time learning and thereby gathering more subjective reality is a valid way to speed things up. Skills take time to master because our mind has to gather ample *mental-copies* and then connect the necessary dots.

The Timeless Mind

We can all recall events from our past. We may do this at will, or because a present time event reminds us. Regardless of the reason, when we do recall a past event it usually feels as if it just happened. The recall feels timeless.

Why is that?

Our *mental-copies* are all here in our mind. Every single one of them is here right now, as you read this book. They can appear to us as something that took place a long time ago because our mind links them one after another, in time sequence. Once we realize that these copies are all present at the same time, then it becomes clear, why any of them can affect us. It's like a huge album of past events organized in time sequence, and we have the album on hand at all times. The event our mind recorded may have been a long time ago, but the *mental-copy* is timeless. The *mental-copies* are the only thing we have about the event or experience.

These *mental-copies* are as timeless as a digital photo. It doesn't matter when we take a digital photo, we can look at it 20 years later and it won't fade. We can look at it (recall it from the computer's memory) time after time and it won't fade or change.

Similarly, our *mental-copies* are also timeless. Once we recall them, they seem like it just happened because they are here with us all the time. Our mind does rank them in time sequence, so if we want to we can look at the time elapsed since the *mental-copies* were taken, but we only perceive the elapsed time if we look at it.

What tends to happen is that something reminds us of an old event which pops in like it just happened. After that, we go "Wow that was a long time ago". Because a split second after the recall, our mind realizes the elapsed time. The vast amount of events since the recalled event is what makes it feel like it was a long time ago. It is understandable then, that since these *mental-copies* are here with us all the time they can affect us. This is why the mind is timeless. We can tell time but the copies are here in the now. We can be accurately aware of elapsed time between events but our *mental-copies* are timeless.

Something that happened a long time ago can have just as much an effect on us, as something that happened yesterday.

Improving Subjective Reality

What does it mean to improve subjective reality? It means to make our own (subjective) reality match the objective (real) reality better. The closer our subjective reality is to the real world the better our understanding will be. Hence the more likely we'll be able to solve problems. Subjective reality is necessary in order for us to deal with objective reality (life and our challenges) and to accurately deal with the world around us. In some cases, it is crucial to be prepared ahead of time for certain activities. Whether we plan to drive, dive, fly, or become an astronaut it is important to have the necessary information ahead of time because of the risks involved. Let's look at how we do that. Communication alone is not enough. One has to gather information and then process it, by thinking about it whenever possible. This allows our mind to connect the necessary dots between the information we have gathered. After school or work wherever and whenever possible. Whatever the information is, we should think with it, and combine it in different ways. Our imagination is the tool to find and connect the dots between the pieces of information we gather. When we have no way to practice the activity directly, it is necessary to use our imagination to connect the dots using whatever we can.

The goal of improving subjective reality is to be prepared as much as possible without having to live through the negative experience we wish to avoid.

If we wish to find out about other people such as fakers, and understand why they do as they do, then we have to communicate with them enough to find out about their subjective reality. Observing them is a form of communication. Then it's time to use creative imagination. We can make believe that we are the other person. Why not? Then we can attempt to see the world through the other person's eyes. The result of this imaginary process will only be as good as the gathered information it is based upon. The more we know about the other person's subjective reality the better we can understand him. What we are doing essentially is making part of his subjective reality become part of our own. The more it becomes our own the more we can work with it. Since this is merely an approximation of his subjective reality, we may not be able to answer everything about him, but we can figure a great many things with certainty. Essentially this is what happens between long-term friends who know each other well. There are countless activities where being able to imagine is irreplaceable.

It's important to differentiate assumption from creative imagination. Just because we use make believe doesn't mean we can't be accurate. If our mind has the necessary details about the other person, we can know with great certainty.

Creative imagination works based on real live information, and can be very accurate when sufficient and error free subjective reality is gathered. Research of all available information is the key here. Only then can we make accurate predictions.

Objective reality exists in space and time, and can be observed and interacted with by all of us and found to be the same. (More or less) It can be touched, seen, smelled, felt, and so forth. Objective reality is something we can all experience.

Subjective reality is quite different. The way we interpret what we observe varies from person to person. Having good subjective reality about a given subject means to have observed and thought of it, and about it, to the degree where we can think with it efficiently. Subjective reality is the basis of our understanding of what was, what is, and what will be. People can think very differently and more or less efficiently about certain things or subjects depending on what they have experienced. Our differences stem mainly from the variations of our subjective realities. These *mental-copies* are accurate to us but since they are copies, they are never as accurate as the real thing. Over time with lots of experience, the copies can be really close to reality.

Therefore, the more *mental-copies* we have of something, the closer we can get to it in our mind. It is important to remember that we don't think with the "real thing", instead we think with our *mental-copies* of it even when we are staring right at "real thing". The subjective reality doesn't have to match the real world exactly, but it is necessary to match it close enough to be workable.

Our logic and creative imagination connects our *mental-copies* as we strive to make sense of things, and thus our mind can fill in the missing details. For example, it is not necessary to walk around a vase that we are looking at to be able to fill in what the other side looks like. After observing the vase from one side we see its general shape, colors, and artwork. From that, without any effort, our imagination automatically fills in the rest. We may assume at that point that the back of the vase has the same continuous shape and artwork.

Unless we go around it, we don't have the full reality of that vase. Just simply being aware of the mechanics of this process can greatly reduce our chances of mistakes. One good analogy to visualize how we gather information is to think of a car and its driver, driving around in a town at night, during a power outage. The driver will only be able to see as far as the headlights illuminate the road ahead. This driver will not have a picture of the whole town available to him. He will only know the parts he drove through already. Similar to this driver, we can only "see" as far ahead in reality as our senses allow us.

Our subjective reality gets built similarly, second by second, picture-by- picture, sound-by-sound, smell-by-smell and so forth. We don't have any other way for gathering information in order to build our subjective reality. It's not like we have a computer plug available to download information to our mind or some telepathic wireless transfer. To improve subjective reality we have to do it piece by piece, line by line, picture by picture, and gather as much information as necessary to have a workable knowledge base.

The more we do this with a given subject the better we get at it. (Practice makes perfect) We can play with the information we gather in our mind any way we want. Our mind is capable of combining our observations in any combination. It can move them around in time and space and can alter it as well.

Our ability to alter what we see is important to be aware of, so that we only do it creatively and with full awareness. In other words, we should not forget that we altered it. If we do, we'll be lost to the degree our subjective deviates from the objective reality. We can alter things in our mind as we play with them in order to solve problems, but when we do this, we should make sure that we don't permanently alter the observed fact. What we observe from the real world should remain intact in our *mental-copies*. Otherwise, we introduce a lie to ourselves, which our mind is quite capable of doing. Later we'll see why this should be avoided. For best results, we should guide our creative imagination to play all we want without altering what we have directly observed. Mistakes, due to altered observation, are very common. When we alter or create our subjective reality based on imagination, it is crucial to keep that in check with reality and observe whether it still matches or not. If it doesn't match, all we should do is acknowledge that it is not matching reality, our mind will do the rest.

Let's see a trivial example having to do with a new driver who wishes to become safe and successful. Observing all the laws and mechanical specifics of driving would be step one in the learning process. Before ever sitting in a vehicle, it is a good idea to think about safety first. Due to the possible dangers involved, it is highly recommended to spend ample time, with the idea of safe driving. Make believe and imagining the future, playing out different scenarios, is extremely useful in preparing for future challenges. Even with a small amount of information, our mind can achieve miracles when we allow it to play.

Statistically new drivers have a higher tendency to get into accidents. Of course, when we talk statistics that doesn't mean everyone has the same tendency. Some of us take the warnings more seriously than others. Chances and frequency of driving also play a role. Some think about the possible scenarios and dangers, and gather more information on the subject thereby increasing their own subjective reality ahead of time. Staying with the example of driving, the idea is not to dwell, or blow the dangers out of proportion, but instead become familiar with these circumstances so that one's own reality is as close to the objective reality of driving as possible. The purpose of this is to be ready and know what to expect, and be prepared. Those who don't take it seriously enough and end up causing trouble, might only learn after they are burdened by the increased insurance cost their recklessness incurred, or worse. Talking to other drivers, reading up on the Internet, are all good sources of information. Information is the "food source" for our subjective reality and our mind.

In order to improve our subjective reality it is not enough to just look, observe, or learn by rote. Genuine interest is crucially important. The higher our interest the more we'll think with the information we have.

Therefore, the keys to improve subjective reality and understanding are interest, and time spent with the activity or subject.

This is why improving subjective reality is not the same as experiencing something. Improving subjective reality means to digest the information we gather well, and utilize it as best we can in order to approximate reality. It can be done ahead of time in order to prepare for something we might experience in the future.

Facts

One dictionary definition of "fact" is:

"A thing that is indisputably the case."

Another dictionary definition is:

"Information used as evidence or as part of a news report or article"

The definition of fact for this book is:

"The undisputed consideration related to matter, space, and time."

This definition emphasizes that we realize facts by thoughts. A fact is not the real universe itself. A fact is some consideration about the physical universe, where the consideration is beyond doubt. This is a subjective process. Why this distinction is necessary will be obvious later as we drill down into how we use facts. A fact can be in the past, present or future, or in any combination of these. (Past and present, or present and future, or past, present, and future) This means that a fact isn't always that way in time or space.

For instance, the fact that "this is the year 2013" is a fact right now, as this book is being written, but soon it will not be. It is a time dependent fact statement. There is time considered in relation to this fact, but most of the time, we do this without thinking about it, or even noticing it, which is why it can be useful to take a closer look at facts.

Something is a fact when the concerned parties agree that it is true. For example, the existence of gravity is a verified fact. As long as we don't fall off the planet we know gravity is at work.

The sun rises and sets each day, is a fact. (Except close to the poles) The sun is hot, is a fact. Ice is cold. Fire burns. These are all elemental facts.

One important detail about facts is the implied conditions, or dependencies of facts. These are frequently unspoken.

Most facts when plainly stated have implicit components or conditions, which we will call dependencies. These are unspoken or unwritten.

For example, "Fire is hot" would be widely accepted by most as a fact, but even that is a relative fact, which depends on the viewpoint. For instance, compared to the estimated 27,000,000°F temperature at the core of the Sun the average fire would seem pretty cold, as even a regular gas welding torch only goes up to around 6330°F. The unspoken component of the fact "Fire is hot" has to do with the word "hot" which is relative. For us (humans) fire is indeed burning hot. The unspoken dependency is that this fact is relative to us (humans). It is a subjective fact most of us know. These examples are purposely trivial and obvious, but when it comes to science, it is important to be acutely aware of what each fact we use really means and depends on. There are many dependencies, which are not expressly spoken or written.

All too often we are expected to accept facts, whether we see the logic in them or not. The safest way to accept facts is to see why and how the fact was initially stated. Then we should follow and understand the dependencies of the fact. Finally, we can arrive at the same conclusion as the person who stated it. It is a good idea to see and verify the thought process of the person who stated the fact. This doesn't mean we should re-verify every dependency of every fact, every time we hear something. For example after hearing the following fact: "The fire can burn through wood doors" we don't have set a wood door on fire, to verify it. Most of us already know the dependencies of this fact. Fire is hot enough to set wood on fire, and therefore it can set anything made of wood on fire. "When exposed long enough" is also an unspoken dependency. Dependency is purposely used instead of expressions such as "implied meaning" to express the structure of priority. Facts rely on their dependencies even when these are unspoken. The necessary description of certain facts can be lengthy. In case of a long description, there may be more dependencies involved. It's a good practice to verify all of these thoroughly at least once. (As necessary) Our mind tends to do this automatically, for most of the dependencies, but an incorrect dependency can slip in occasionally since they are often unspoken. This can happen due to outside influence or pressure by time. Pressure can force us to skip the verification process. A trusted source can bypass our verification process also. An influential authority can also cause us to bypass or override our verification process.

When someone is being hard-sold on a car or house, the clever salesperson has the routine down already, to keep throwing sentences after sentences at the potential buyer. If the buyer appears bogged down, the pushy sales man will immediately start some other line, and keep explaining the same thing from different angles. The salesperson is not likely to mention the fact that the credit the buyer receives for his current vehicle is way under fair market value. This is a fact the salesperson is keen to hide. This is the perfect strategy to get someone to skip over important facts.

The salesperson states something as a fact: "This car is perfect for you". After the buyer tells the seller his requirement, it is unlikely that the salesperson would say something like "Considering your requirements, you should go next door, to the other company, and buy a car from them instead". Such honesty is rare but it does happen. Therefore, in reality the sentence, "This car is perfect for you," really means, "This car is perfect for you from what we have to offer." The salesperson omits the "from what we have to offer" part, which becomes a hidden dependency. This would not throw off the determined buyer, who wants to test more cars before buying. Others, on the other hand, might be influenced enough to buy. All it takes is to miss the omitted dependency.

The unspoken dependencies frequently allow false-information to enter our thoughts. It is a form of omitted information. People who manipulate others exploit this frequently. With practice, and attention, we can become more aware of the unspoken dependencies of facts. This way we can reduce false-information and increase our overall awareness. Then we can reduce the effects of manipulation and influence on ourselves. We can also point these out to others around us. Just by understanding the logic of dependencies most of the work is complete. Then each day we can just set a simple daily goal to find an intentionally omitted dependency. This can become a useful, but effortless habit. This can increase our curiosity, about facts, so that we can see all the details related to them. Curiosity combined with the willingness to verify what we hear or read, can make a positive impact. It is a rewarding feeling to be able to spot and resist such manipulation and negative influence.

Elemental an Complex Facts

Elemental-fact is defined (for this book) as an observed rule of interaction in the physical universe that is always true. *Elemental-facts* are those that we cannot subdivide into smaller units of facts, as well as those that we chose not to subdivide any further for practical reasons. *Elemental-facts* are empirical, and thus we can observe them to be true for ourselves, even if sometimes that observation is more involved. For example, the *elemental-fact* that water boils at 212°F we can observe easily at home with stove, pot, water, and a thermometer. The value of π (PI) is also an *elemental-fact*, and we can observe it for ourselves, that the ratio of a circles circumference to its diameter is in fact π, that is 3.141592... For everyday life, *elemental-facts* are stable observations that have characteristics that do not change. *Elemental-facts* also don't depend on time. In other words, it doesn't matter when we state it, or examine it. It must be true; otherwise, it is not an *elemental-fact*. Another *elemental-fact* is the speed of light in vacuum, which is more involved to observe and prove, but can't be further dissected (at least for now). The existence of heat is an *elemental-fact*. Heat does exist without a doubt. All matter is capable of absorbing heat when it is applied. The ability of matter to interact across distances is also a fact. Anyone can observe this when using a cell phone. The cell phone transmits voice wirelessly back and forth to another phone. This can be seen as an *elemental-fact,* if one doesn't chose to know the underlying physical phenomena called electromagnetic waves, which enable cell phones to communicate with the cell phone tower. *Elemental-facts* therefore are subjective. It depends on the individual's knowledge about the universe.

People can look at a pool of water and have an understanding of water as they observe and experience it directly. If you throw in a rock, it will sink. Put a toy boat on the water it will float. Leave an iron tool in, it will rust. Such facts are sufficient for most people and for them these are *elemental-facts*. These we can observe and know it is true. Scientist had to dig deeper and come up with other *elemental-facts* invisible to the human eye such as atoms and molecules and their interactions. In life, there are many devices and phenomena, which we can understand on different levels. Depending on how deep we decide to go with our knowledge, we need different levels of *elemental-facts*. The level at which we settle is a matter of personal choice. This means that we don't have to go too deep to be able to have *elemental-facts*. Anything we can observe to be true we can use as an *elemental-fact*. After declaring our *elemental-facts*, we can build solid theories on them. If we have observed it to be true and verified it, then it is true. Just because someone else dug deeper doesn't make our *elemental-fact* less valid at all. Our mind creates *elemental-facts* automatically as we experience new things.

Complex-facts are those facts that require, and depend on two or more other facts. Thus, *complex-facts* always contain at least two or more other facts. These component facts can be either *elemental-fact(s)*, or they can be other *complex-facts*. The glue that connects these component facts is logic. *Complex-facts* can be built and expressed using any number or combination of other facts. In this book, facts are either elemental or complex. If a fact is not possible or unpractical, to be broken down further, then it is an *elemental-fact*. Everything else is a complex fact. A simple example of a *complex-fact* would be a suspect who is dismissed due to a solid alibi such as a security video showing him at a mall in a distant city at the time in question. The component facts are these:

- The time of the incident
- The place of the incident
- The time the suspect was seen in the video
- The place where the video was taken
- The authenticity of the video

Spotting Fakers

- A person cannot be at two places at the same time

In the court of law, logic and facts are superior, governing rules. When the suspect is dismissed based on the evidence combined with logic then a *complex-fact* is being realized and accepted by the court. That *complex-fact* is as follows: The time and place of the incident is known. The authentic video showing the suspect being 300 miles away at the time of the incident, combined with the logic that he could not have been at two places at the same time, results that the *complex-fact* can be stated: "He was not involved in the incident".

This is a trivial example, but shows the breakdown of a *complex-fact*. In such a case, we might not even be aware of all components, but our mind does consider the facts automatically to reach the conclusion. Facts such as "He can't be at two places at the same time" are self-evident and we don't have to be reminded of it. Such facts are often unspoken. Some *complex-facts* can't be directly observed. In such case, we can rely on other facts (elemental or complex). Combining these with logic, we can create and state a *complex-fact*, and see it to be true. We use this frequently to figure out what we can't observe.

The important rule about *complex-facts* is that if all the component facts and logic is verified and true, the *complex-fact* will also be true, regardless of size and complexity.

Consequently, if any of the component facts or logic is false then the *complex-fact* will most certainly be false also.

This is a very important factor because it shows how a simple error can bring down (make false) impressive looking and long theories.

In the court of law as well as in the scientific areas, *complex-facts* are readily accepted. (Most of the time) As long as all of the components facts and logic is undisputed, they should be. To disprove a *complex-fact*, such as a new discovery, it is necessary to find and show the incorrect fact(s) or logic.

It is common for certain people to criticize the findings of others, even if they can't find a hole on the presented facts and logic. This is non-creative criticism. Being able to see the dependencies and the importance of facts can help us spot and ignore such non-factual sources.

Logic

A dictionary definition for logic is:

Reasoning conducted or assessed according to strict principles of validity: "experience is a better guide to this than deductive logic."

The definition of logic for this book is:

"Logic is prediction"

This definition doesn't make the dictionary definition false in any way. It merely breaks logic down to its smallest building block. Being able to see logic in everyday life as an action of prediction may not be so obvious without an exploration into the nature of logic.

We live in a universe where events happen sequentially.

Does that mean that the universe itself is sequential? Not necessarily but that's an entirely different subject. One thing happens, then another, then another, so on so forth just like watching a movie. That is how we perceive reality. A movie in a theater that is. Not in your home where you can pause, rewind, and fast-forward or even edit. The reality of the universe is that it moves sequentially forever forward, whether we like it or not.

If we make a fatal mistake, truly undoing it is not possible. Making up for it is not the same as undoing. Undoing would be to travel back in time and do it differently hoping for a better outcome. This isn't even remotely possible with our current technology and understanding of the universe.

Any video recorder records a film by capturing and storing a bunch of individual pictures one after another, in an exact sequence. When we play a video, the pictures are given back to us in the same exact sequence as captured. This method works well enough for us, as we perceive the world as a sequence of events also, and try to make sense of it in a sequential manner. That is one picture (or event) after another, then another, then another and so forth.

For our purpose, it's not necessary to make assumptions or statements whether the universe itself is sequential in nature or not. The fact that we perceive the universe in being that way is enough. After 9:00am comes 9:01am then 9:02am and so forth never vice versa. We perceive the universe based on sequential observations. Logic is the ability to predict the state of something based on previous observations.

Let's see this in a different way: logic takes advantage of the fact that the universe's hands are bound by rules. These rules allow us to: (1) Make observations by creating *mental-copies* of the universe around. (2) Based on the *mental-copies* attempt to make a prediction. As we learn about or observe facts, it is still an observation. We are constantly observing the universe around us (recording). At least the portion of it that is close enough to be observed. We can extend our range of view using telescopes, microscopes and remote cameras of course, but that's still just observation of some display or lens that is close enough for us to see.

Logic is to take some information, and based on that information, make an accurate prediction that is not expected to fail. The "accurate" part is important. If there is any doubt, whatsoever, in the outcome then that prediction doesn't qualify as logic, instead it will be a guess. Logic is only possible, and based on, the sequential nature of our observations and the underlying sequential nature of our universe (as we know it).

Spotting Fakers

Here is an example logical thought: "If I touch the pot my finger will get burned".

If someone is thinking this and this thought is voiced in his mind that doesn't mean that this is the only thought his mind has about the issue. Voiced thoughts refer to the thoughts of our internal dialogue. Voiced thoughts are silent and unspoken but in our mind, we perceive these as spoken. Consequently, the ones that we say aloud are spoken thoughts. Important to note, that voiced thoughts can exist without speaking. There are other component thoughts present, even though these are not voiced. These component thoughts are not as obvious as we don't voice them in our mind individually and they only take a split second each.

Let's look at the possible component thoughts prior to "If I touch the pot my finger will get burned":

- Remember setting fire
- Remember fire being on
- Remember pot over fire
- Realize time elapsed since fire was set
- Observing fire now
- Observing pot over fire now

The following may sound like a description of a program, but in reality, this is in fact the sequence of thoughts:

Since the fire was set <u>and</u> the pot was set over the fire <u>and</u> sufficient time passed since <u>and</u> the fire is still on <u>and</u> the pot is still over the fire; the conclusion is that the pot is now hot enough to burn the finger. The result is the final thought: "If I touch the pot my finger will get burned."

The fact that we don't voice these component thoughts, doesn't mean they were not there in the first place.

Consider it this way:

Someone can just look at a pot, know all these facts, and not touch it without a single thought voiced in their mind. Some thoughts are voiced while others aren't.

Fortunately, not all thoughts have to be voiced. Voiced thoughts are much slower. The mind does its native (voiceless) thinking very efficiently. Why we voice anything in the first place when no one is around to speak to, is a subject that's covered later. The voiceless split second thoughts of the mind are *native-thoughts*, covered later in the *Native Thoughts* chapter.

The prediction is infallible only when all facts that our knowledge is based on are true. For instance one could think the pot is hot, not knowing that someone else had turned off the fire (allowing it enough time too cool off) and just restarted it five seconds ago. The logic of the mind is extremely accurate but if the information we work with, is incorrect, then our final judgment can betray us and we can make mistakes. Without assumption, this is how we could think of the same scenario:

"Unless someone else kept the fire off while I was not here, if I touch the pot my finger will get burned." This sounds wordy and long but in reality, our mind considers much more factors in less than a second.

A simpler but assumption free way to think of it is:

"It is very likely that, if I touch the pot my finger will get burned."

To us this may just register as the lack of urge to touch the pot without any voiced thought.

In this elemental view, logic and reasoning is the action of predicting the state or condition of things based on other observations. The word "things" is used here intentionally so the sentence doesn't get too long. "Things" means any physical object or subjective thought. For logic to exist there has to be a unidirectional sequence of events. The goal of logic is to be able to predict future position and/or condition of things in space and/or time. This is what logic looks like in its most elemental structure. We do this all the time and depending on the circumstances, and the complexity of the given scene, we use information in an attempt to compute and predict the future. This is true even when we make logical deductions about past events. The events of the past are also in a relative sequence to one another. Past events also had to follow the sequence. Thus, we can apply logic to past events. Sequence applies to concepts of events as well.

Without sequence, there would be no logic.

"Information" refers to one or more of the following:

-   A universal constant
-   An elemental fact
-   A complex fact
-   Something simply stated to be (whether right or wrong)

Information can be excessive comprising many of each of the above. Information can be any observation in present time or recalled from the past. Information can also be observations of others, communicated to us. Information covers anything that we can describe, communicate, or conceptualize.

Depending on the subject, study, or problem we address the logic and facts can be involved. It may seem that this can get overwhelming and hard to tell when and what we should verify. Fortunately, as mentioned already our mind can tell us when something is not adding up. We feel it. All we have to do is listen to the split second message from our mind, which tries to tell us. Once we get that message or feeling, we shouldn't go any further in any subject. Ever. Instead we should find what's not adding up.

Any information can be comprised of many other information units. Life is full of false-information and traps that can cause us to make mistakes if we accept them.

Let's take "Fire is hot" one more time as an example:

There is the observed level of truth to this as we grow up and have a picture-like concept of what hot is. The closer we get the warmer it feels. If we get too close, it gets unpleasant, or painful.

When we are cold and sit next to a campfire, our mind will access the *mental-copies* with the pleasant nature of the fire. We see the fire and our *body-sensors* inform us of the presence of heat, but not too much heat. The right amount of heat. We also know we are sitting safely by the fire. Based on these facts and observations we predict that what we can expect is a pleasant experience. If we stand up and get too close, followed by losing our balance, our mind will access different *mental-copies* and concepts about the fire very quickly. These are going to project a much less pleasant experience. Signals from our *body-sensors* plus the increased intensity of heat, combined with our body's position leaning over the fire would be some of the factors our mind would be using. The sequential nature of the universe combined with gravity, predict that we are likely to end up inside the fire unless we do something quickly. Gravity is not likely to give us a break with such predicament, it's better to grab onto something. All this takes place on the *native-thought* level. This recalling is automatic, and below our level of voiced awareness.

Our *native-thoughts* and thereby logic relies on the sequential nature of the universe. That's how we predict.

Of course, we use other laws of the physical universe for our predictions also, and create facts from these based on our observations. The most elemental component is the sequential logic, which is the glue that holds it all together.

This is logic at its fundamental.

Knowing

Knowing is the ability to predict something using logic and having accepted that prediction to be true. Accepting means to have no more intention to verify the facts involved in the knowledge. Which means we don't want to compare the facts to reality anymore because we consider that the facts match reality close enough. This does *not* mean that they are in fact close enough. It just means we have considered it does.

Knowing is not the same as logic. Logic is the "glue" that allows for prediction. Logic is based on information (facts). Knowing is the end-result of having made a prediction on a given item or subject, and accepted it as valid. That is, we consider the prediction verified. Knowing is not the same as having beyond a doubt verified that the prediction is true. Knowing simply means we have accepted it to be true.

For example the sentence:

"Fire is hot" is a known fact for those who have experienced it, and thus have the ability to recognize it and predict things based on this knowledge. "Fire is hot" only becomes a known fact once someone accepts that to be the unquestioned truth.

It is possible to prioritize facts, logic, prediction, and knowing. At the first layer, there are facts (elemental or complex). The next layer is logic, which depends on the first layer (facts). Logic requires facts as a foundation. The third layer is the prediction, the product of logic, which we apply to facts. The final layer is "known".

Known is the state of having accepted the prediction of the third layer to be true.

139

Here is a list of the four layers:

4. Known

3. Prediction

2. Logic

1. Facts

The dependency is important.

#1 is the most important which will affect all the layers above it. Incorrect facts (containing false-information) will cause the other layers to be false. Facts are the foundation of knowledge.

Knowledge is the collection of many different known items. Each item is a piece of accepted prediction. As we practice any skill, we use these individual accepted predictions without even realizing that we are predicting various events. With practice, we can spot this in the activities around us. In turn, this allows us to spot and prevent mistakes easier. Let's look at a trivial example of pushing the call button in an elevator on the 5th floor of a fifty-story building. This elevator has up and down arrows that light up to indicate the direction. When the elevator arrives, we can use the lit arrow as a "fact" to decide whether to get in or not. If we are heading down to the lobby, and the up arrow is lit, we predict using logic that if we get into the elevator it will not be going to where we want. We can also predict that if we do get in, the elevator will eventually go down, but first it will go up. We can also predict that if we opt to wait for another elevator door to open with the down arrow lit then this will take less time to get to the lobby. If time is important, we can predict that this option is a better choice for the fastest way to the lobby. This is the underlying logic in the knowledge that "It is faster to get somewhere with an elevator if we wait for the one with an arrow that matches our desired direction." Such trivial decision already involves multiple known items and therefore elemental predictions. Let's see some of the components. When the down arrow is lit, the prediction is that the elevator will be descending. Up arrow predicts ascending. We compare the product of the prediction to our desired destination. Our desired destination "to descend to the lobby" is a given which can be considered a fact. Another prediction is that the elevator will eventually descend regardless of the arrow because it can only go so high before it has to descend again. Other possible predictions include that if we get in despite of up arrow being lit on the 5th floor of a 50 story building the elevator may stop an unpredictable amount of times to allow other passengers in. Since we don't know which floors have been already pressed inside the elevator and we don't know who else is waiting on upper floors to go even higher this is totally unpredictable as to how long it will take. The worst-case scenario in an extremely busy building would be that it stops 90 times before we even get back down to the 5th floor. (If it has to stop at every floor up and down) The best-case scenario is that it only goes to the 6th floor before it is allowed to descend again. Huge difference. Our mood and schedule are also factors we consider in the prediction process. If we are not in a hurry then we might just be game enough to see some new faces and try our luck going up. Then our considerations for others may come into play also. If the building is so busy, the elevator is likely to be stuffed. Are we going to hog up the space? Space in an elevator is finite. (Another fact) It's a personal choice. If we are on a tight schedule then it's best to wait until a down arrow lights up. The whole decision logic may change if we do the same on the 49th floor of the same building. The worst that can happen if we get in on up arrow is that we go up one floor before we descend. This trivial example demonstrates how seemingly simple decisions can have many component predictions, which we use without even having to think about it. That's actually figural speech because we do have to think about it, but we are doing that subconsciously using *native-thoughts*. We might voice some of those component thoughts out occasionally but our mind can navigate us quite reliably without us having to be aware of it. All the above facts and known items as a collection could be called the knowledge of riding elevators.

Whether we take an IQ test or attend a sports event, we can explain the involved logic in terms of predictions. If we dissect our actions and decisions the same way as we did in the elevator example, then we can discover the component facts and logic. Then we can see the important role sequence plays in our logic.

Knowing is the consideration of a prediction to be true. Keep in mind that knowledge is a combination of many "known" items. Our mind also uses known items as a fact, for building new predictions. In reality, our mind has many lengthy chains of these by the time we learn our native language. The interconnections and complexity of these can be time consuming to analyze, but we are not interested in doing that. What we are interested in is that these are chains and false-information is the weak link.

It doesn't matter how long a chain is. If there is false-information involved anywhere in the chain, our mind is capable to indicate to us "Something is not adding up". If we don't ignore that quick indication, then we can look for and find the weak link.

False-Known

It is possible for us to accept something to be true and known, without sufficient verification. The lack of verification opens up the possibility of false knowledge. Accepting a logical prediction, which is based on false-information, causes us to have a *false-known*. A *false-known* therefore is something we think of as certain but it is incorrect. People we trust can sometimes give us incorrect facts, or incorrect knowledge, and due to the trust we have in them we accept these to be true. Right there and then, the moment it's accepted, it becomes our *false-known*. This isn't necessarily intentional. They may and usually do have the best of intentions, and to their best knowledge what they have presented was true. Which means they had that as a *false-known* also. False-information, and a *false-known*, can pass from one person to another this way, via the power of trust. What *is* trust anyway?

In this case, it goes something like this:
"If he/she said so, then it must be true".

Trust then is the consideration that the person, who is providing the information, must be beyond any doubt and correct at all times.

Could that trust be a *false-known*? Let's see how we accept someone as trusted. When someone is kind or helpful most of the time, we consider them trustworthy. Also, when we can't find a flaw in someone else's reasoning we tend to accept that person as a trustworthy source of information and knowledge. Once we do that, we are likely to accept everything else from the trusted source and skip the verification. This is not the same as them always being correct. Due to the large amount of false-information and *false-known* that is present throughout society this does not seem like a bullet proof approach. Even the most experienced and

143

scientific person may have some false-information slip in occasionally, especially on subjects that are not part of their main field of expertise. If that is so then no one is beyond doubt. In which case instead of trust, we should consider people who seem to be always right, as *people who are mostly right*, but information or theories from them is still subject to be verified. Verification of all information, all the time, is crucially important. This isn't a bothersome task to be added to current duties. It's simply a change of view. Where we respectfully acknowledge the opinions, information, and conclusions of others, but don't just accept it unconditionally, instead note to ourselves that it should be verified. If we pay attention to this at first daily, then over time, this develops into an effortless habit. Just by instructing our mind *not* to store the information as complete and accepted, we cause it to be open for future verification. It's like filing the information into a "to be verified" folder in our mind as opposed to putting it into the "known, case closed" file. Then our mind will work on figuring out the details, inherently, as that is what it's meant to do. The mind automatically connects unconnected dots. By doing this, over time, we can encourage our curiosity and dot connecting abilities which allowed us to go, from zero to walking and talking in a short time.

There is nothing harmful about false-information, if we spot it. Only when we accept the false-information as true we get in trouble and have a *false-known*. A *false-known* that someone else tells us, is essentially just false-information. It only becomes our *false-known* if and when we accept it as true. If we spot someone else's *false-known* we should label it as "not sure". Then we can't get into trouble because of the information.

For example, imagine that during a conversation some tells you something, which makes you realize that this person thinks all tigers are the same color. When you spot that *false-known*, you may want to help this person by letting him know. We should do this gently for best results. If we care about the person (which we do), instead of telling him "Tigers aren't all the same color", first we should spot the false-information that makes it a *false-known*. Perhaps he never saw a white tiger before, or if he did, he might have thought it was a different species. Then we can really help the person in a casual and indirect style such as "I kind of remember seeing white tigers once in a TV show". Then he might say, "You mean not all tigers are orange and black? Wow, I thought they were all the same color. What was the show you watched? You know I just realized something. Maybe orange black tigers aren't all the same either. Maybe they have all sorts of differences. Wow! That blew my mind. Man it's fun talking to you." such gentleness can make a world of difference. If we tell this person, that he has a *false-known* to him it comes off as "You are wrong!" which the mind tends to perceive as "Damn, I'm broken, see I made a mistake again". Which is not true. As mentioned before there is nothing wrong with his logic. He just has missing information because he had never seen a white tiger. False-information of the missing kind caused him to assume and accept that tigers are all the same color. You gently bring it to his attention that you have seen a white tiger before. What you are doing is filling in the missing information. He can deal with this very nicely without him thinking there is something wrong with his mind. With the false-information handled, he comes to realize and handle his *false-known*. He is doing that for himself and he will feel good about it. You will see his face light up guaranteed.

This happens to be another good spot to say what a faker would do about the tiger *false-known*. Just because they don't care about others doesn't mean they can't be logical and cunning. Indeed when it comes to indirectly putting others down, they are the masters. They can spot a *false-known* of someone else which makes a perfect opportunity to put that person down. A faker would most likely say something like "You are wrong! Tigers are not all the same color. I know because I saw them in Las Vegas". This is a potentially what the other person will think: "I'm wrong". "Again". "See I'm slow". "Also I'm unlucky because I've never been to Las Vegas before, while this guy has been traveling all over the place". The result is that he is not going to feel good because he will not see that his problem is as simple as missing information.

This approach is important when we address ourselves also. Whenever we ask ourselves a question, it is similar to addressing someone else. When we finally solve a persistent problem, often we don't spot the false-information that was the culprit all along. Then because it took so long for us to figure, we may have a bittersweet feeling. On one hand, we feel good for succeeding finally, on the other hand not so good, because it took so long. If we see someone else having solved that problem faster (as it can happen in schools a lot) that is even more devastating because we may end up thinking the other person was smarter. In reality, what happened is that we needed longer time *strictly* because of false-information. This is why it is important to spot the false-information even after success. Every single time. This automatically removes any negative considerations about our abilities. "Oh, it was the false-information that held me up, not me!" "I'm fine." "I'm good." Spotting the false-information makes a huge difference.

Another important aspect about *false-known* is that one can't see it. The *false-known* is not visible to the individual who has it. If it were, then it would immediately stop being a *false-known*. Suspecting some knowledge to be a *false-known* is not enough by itself for handling it. We handle it by finding the false-information it is based on.

One amazing fact about a *false-known* and human logic is that logic appears to be incapable of error. It's a common misconception among people to think that their logic just failed them. Behind most (in not all) of our errors and mistakes, false-information is the culprit. When it seems like logic is at fault, extended research into the details always turned up some false-information as the root cause. Based on this observed fact, the logical explanation would be that the mind has redundant self-checking and is indeed fail-safe. We may be referring to human errors as brain-farts, brain-cramps, or brains-shorts. Regardless of the name, the cause of human mental failure is not what it seems. The mind is not failing. Therefore, the common expression "it is human to err", couldn't be further from the truth. Accepting false-information as true is what causes us to make mistakes and err. If we better the overall quality of our knowledge base, as a society, and individually, our mistakes will reduce radically. An additional observation, which supports the theory of our logic being infallible, is that we tend to get upset when we realize we made a mistake. We ask:

"What? I made a mistake? How on Earth is that possible?"

No one likes to be wrong. It would make sense that we inherently know that our mind is not supposed to make mistakes. When we make a mistake anyhow, it is indeed puzzling. We may feel humiliation, anger, insecurity, and just about any kind of negative emotion. It should be a relief to know there is nothing wrong with the logic of our mind. The only thing required is a good housecleaning of false-information, and habits that can assure we don't get more false-information in the future. Sometimes we discover a *false-known* by accident and handle it. That means we spot the false-information that was causing the *false-known* and correct it. In such a case, our mind automatically sorts out any other dependent knowledge. Our mind's marvelous linking system does this automatically. Thus, the only tricky part of handling a *false-known* is finding it.

A *false-known* can always be traced back to its false-information content. Keep in mind that omitted and missing information is also a form of false-information. Therefore, missing information can also cause a *false-known*. How? When we are faced with omitted or missing information, we are subject to assumption.

What *is* an assumption anyway?

Assumption itself is a *false-known* also. The content of this *false-known* is that in the presence of omitted or missing information, it's acceptable to guess (make up something), and accept the result of the guess as known and true. That is the breakdown of how assumptions can get us in trouble. A guess is a guess. It is to be treated as a guess, and thought of as a guess, until it's proven to be true. Only then should it be accepted as known. An assumption can turn out to be a *false-known*. Therefore, it is

147

subject to the same rule of being passed from one person to another. We can pick up the habit of assumption just by being around people who assume frequently.

A note about the term *false-known*:

The word misconception is purposely not used.

Misconception means: *"A view or opinion that is incorrect due to faulty information or thinking."*

A *false-known* is strictly caused by faulty information. Thinking involves logic, which is not what's at fault. Faulty thinking is always caused by false-information, thus a new term was used, to explicitly express the source of trouble.

Deriving Facts

The facts underlying certain phenomena are not always observable even by direct experiments. It may be necessary to rely on indirect observations. In such case, we can gather enough observed facts and use these to derive (figure out) the unknown fact. The unknown fact in such case is something we can't directly observe. We'll call this a *derived-fact*. In simple terms when we figure out the unknown thing, based on more than one known things then we realize a *derived-fact*.

A *derived-fact* is essentially a *complex-fact*. It's the same thing. It's only named differently in this chapter to indicate the required function for the realization of the fact.

It is important to note that logic is only infallible in the mind. Once we put a fact (complex or elemental) into words, the logic is also subject to verification. This is because the words and sentences that we use to express the logic, can introduce false-information also.

Spoken, written, or otherwise expressed logic is fallible, and subject for verification.

Astronomy and particle physics offer good examples for *derived-facts*. In astronomy, the limitation is that we cannot travel to distant solar systems and galaxies to observe them. In particle physics, the problem is that even with advanced microscopes we can only observe whole atoms or molecules. Direct observation of particles at the subatomic level is not yet possible. This doesn't mean that we don't have a way to find out about the nature and underlying facts of sub atomic particles, distant planets, star systems and galaxies. Scientists spend enormous amounts of time, money, and effort every day using space telescopes, particle accelerators, and other complex and expensive toys, to gather as much data about these uncharted realms as possible. The information they collect is analyzed, and verified. Using these, they derive the facts we can't see.

Fortunately, we all have the ability to use and create *derived-facts*. Finding patterns and connecting dots is a natural ability to all of us. If we didn't have this ability, we could never learn a language in the first place. The fact that we learned a language is the proof that our mind has the ability to connect dots extremely well. We don't need an expert to tell us, it's obvious. The fact that we think, walk, and talk is more than enough proof to this. Whenever we figure out something, we did not observe directly then we realize a *derived-fact*.

Another good example is the person in the chapter *Elemental Facts* who had an alibi. We can't observe directly whether he was at the scene of the incident or not because it is in the past. No witnesses or video recording is available. This means we need to figure out something that we can't directly observe. The incident happened already. There is no way to go back in time to observe if he was involved. When we accept his alibi, we are realizing a *derived-fact*. Based on the other facts we figure out the unknown. The unknown is "Was he involved or not?" The known facts allow us to figure out that he wasn't there.

Derived-facts are a necessary part of all sciences and knowledge.

A fact based on assumption, is not the same as a *derived-fact*. A properly *derived-fact* is based on research, and when the research is correctly done with attention to possible false-information sources, the results are accurate. Only false-information, *false-known*, or insufficient research can make a *derived-fact* false. Our logic is never at fault. The following example will show just how accurate *derived-facts* can be with proper research.

The first person to propose and write down an atomic theory was the Greek philosopher Democritus. (Somewhere around 460 to 370 BC) More than 2000 years later, in 1803 John Dalton presented the first truly scientific theory of the atom. These were all theories *derived-from* experiments and observations yet they were both correct. It wasn't until the 1950's that Erwin Wilhelm Müller first experimentally observed individual atoms using his invention called the Field Ion Microscope. In other words, he was the first man to "observe" individual atoms. This confirmed beyond a doubt, that atoms exist. "Seeing is believing" was not necessary, for all those before him to accurately predict the existence and many behaviors of atoms. They were able to derive the facts from observed data and experiments without "seeing". There are numerous examples in life and science, which can be used to demonstrate that *derived-facts* can be very accurate when the research is done right. Those *derived-facts* which prove to be incorrect, can be traced back to the false-information or insufficient research as the cause for failure.

Assumptions

As mentioned before an assumption is essentially a guess which we have accepted as true. A guess is harmless as long as we know that it is a guess, and we continue to think of it that way, until proven to be true, or false. Whenever we guess, it means that we didn't have enough information available to know. It can also mean that some false-information is causing us to be unsure. Without sufficient subjective reality, we tend to rely on assumptions.

Since assumptions are accepted guesses, we increase the chances of false-information entering our subjective reality. For true knowledge in any subject it is required that we have little or no false-information present in our subjective reality. Therefore assumptions are not helpful and best be avoided

It is important to note that an assumption is not always wrong. An assumption is like flipping a coin to decide which way something is. There are usually some known factors present when we make an assumption. It is not a black and white type of deal. Depending on how much we know versus how much we don't know the chances of the assumption being correct will vary. If most of the information we know is accurate related to the assumption, and only a few things are missing, the chances of the assumption being correct are relatively high. On the other hand if we are assuming something, with little or no valid information, the assumption has very little chances of being correct. Whether an assumption is right or wrong, is best viewed in terms of statistics.

Each time we assume something, we introduce a certain chance to be wrong. Assumptions that build on one another also accumulate. For instance, if we create a theory based on three consecutive assumptions, each having 50% chance of being right, and 50% chance of being wrong, then the accumulated chances of being correct is reduced to 12.5%. That means we would have an 87.5% chance that the theory is incorrect. In the case of three consecutive assumptions, with a 10% chance of each being right, the overall chances of the theory being correct would reduce to 0.1%. That means only one out of a thousand chances that the theory is right. These look like gamble. Therefore assumptions are not well fit for science and accurate knowledge. If our mind is equipped with infallible logic, there is no reason why we should make inaccurate and unnecessary assumptions. It is much safer to research the information we require, or find and fix the false-information. That way, we can know, as opposed to having to guess.

Assumptions are accepted guesses that are more often false because we usually have more than one on a given subject. Therefore, an assumption we accept as true is more likely to be a *false-known,* then a known. Whether the assumption was right or wrong is not known to us, until we try to verify it. The problem is that the mind doesn't do that automatically unless we develop a habit for that. It is necessary to intentionally direct our attention to finding false-information which in turn reveals and handles *false-known* as well as false assumptions. If we are forced to make a guess, it is important to note to ourselves, that we are not 100% sure. Also a good idea to note to ourselves, that the guessed subject should be researched as soon as possible. It is never recommended to accept a guess as true since that makes it into an assumption.

Assumptions can also lead to upsets as well as self-invalidation. When people show up to their first work place, their reality about the place is likely to be based on assumptions. As they spend time in the new environment, their subjective reality about it is not only built, but corrected, as they replace their previous assumptions with real observations and facts. In this process, it is common for us to find certain things not to be as expected. This can cause severe upsets and reduced trust in our ability to know things. It is possible for some assumptions to stick around even when contradicting facts are present. Since a *false-known* is not visible to us, assumptions when accepted as true can also hide from our awareness. In such case we have a tendency to dismiss or ignore the contradicting facts even if we are looking right at it.

In society it is easy to develop an attitude of "I must know" which is a form of rushed knowledge. This is understandable, since in today's fast paced and challenging environments, we often feel like we don't have the luxury of time to research everything. Many work places and schools can make us feel like "It is *not ok* to not know".

As a result we often force ourselves into assumptions. We may even pretend to know stuff, that's how socially unacceptable we may perceive "not knowing." How to deal with what we don't know is crucial but so simple that it's bound to be overlooked.

It is important to realize that, it's ok not to know certain things. Time is finite, and we can only learn so many skills and information in a given time. It is inevitable, that we don't know every single thing there is to know so we specialize in certain fields. When we encounter something we don't know, we should be able to acknowledge that. Immediately we can file it in our mind as "not yet known", or "to be found out", or "to be verified." Our mind files and categorizes all that we learn anyway. All we need in our mind is one more category (or folder) labelled: "not yet known." We already have tons of categories that our mind uses. Gardening, mathematics, chemistry, movies, songs, astronomy, restaurants, grocery stores, schools, school mates, cooking, workplace are just a few categories to give some example. We can think of these as folders, in which the mind files new experiences and knowledge. Every new experience and knowledge gets filed in the appropriate folder, without us having to pay attention to it. Due to the already mentioned influences, even when we have an "unknown" folder, we don't like to file stuff there. We would rather guess about it, and file it as a "known" in some other folder. The name of the folder can be phrased in many different ways. Any label will do, as long as it represents, and acknowledges the fact, that we don't know what it is. It takes practice to break this habit. (Just like any other habit) Just by realizing that we haven't

155

been filing unknown things where they belong, in the "unknown" folder, can jump start the process of kicking the habit. It is an important step in getting rid of false-information. Also just by getting into the habit of acknowledging that we don't know something, we can eliminate gathering further false-information. The more we do this the easier it gets. As we improve our skills in any area it's important to know more and more. Equally important is to identify, what *it* is, that we don't know. To know something first we have to realize that we don't yet know. Then and only then can we learn.

We can't learn something we think we already know.

Assumptions can be right sometimes. This may look like a good thing on the surface, but let's take a look at what happens when an assumption happens to be correct by chance. If for example someone has low expectations of a new work place, and find the place is much nicer than he assumed, he will be really happy. This may seem to balance out those times when he felt let down by a false assumption. Unfortunately this isn't the case. When we assume and the assumption turns out to be correct, what we do to our mind is strengthen the *false-known* according to which, assumptions are a good way to deal with life. This in turn makes assumptions acceptable instead of true knowledge. Assumptions will be known as something which sometimes works and are useful. (Better than nothing) Then we begin to think that knowledge is a hit and miss endeavour. If we assume often, knowledge and life will begin to seem like a big gamble. As a result, the idea of being able to know things for certain may disappear. This is how false assumptions make us uncertain.

Knowing with certainty is possible for everyone, but in order to achieve that we should eliminate assumptions and false-information. When we don't know, we should acknowledge that fact to ourselves and then research it as soon as possible. For little things, this can be as simple as an Internet search (like looking up a phrase of theory), for bigger issues, where a skill is required, it may take years of research and learning. If that is too much effort, because we have other obligations, then we better just admit to ourselves, that we don't know and have no intention to learn it for the time being. This way we can stay away from assumptions and learn at a later time if we wish.

Learning by Rote

Learning by rote, means to store and be able to later remember, and thus recite information, without actually understanding the contents of that information. This happens when learning is forced but our interest is elsewhere.

When forced to learn something, we only communicate on that subject as much as necessary, and then be on our way to do what we really want to. This is a minimal form of learning and not very useful. It doesn't improve our reality on the subject efficiently. We record but connect little or no dots.

Subjective reality only becomes useful when we take in the information, then actively think and play with it. There's nothing fundamentally wrong with this, it just shows how dedicated we are to the things that interest us.

Learning by rote, is an attempt to fool the enforcer of learning, into thinking that the subject is learned. Based on the rules of subjective reality, it isn't difficult to understand why there's little use for such imitated knowledge.

If we don't play around with the information in our mind, then the necessary logical links between all the pieces of the information won't be created. The dots won't be connected and without these links, the mind can't call on these subjects as necessary. The information will be there, but only as one big continuous recording of sounds, images, sentences, and phrases. The individual sound, image, and concept fragments won't be readily accessible to think with.

This calls for a quick look at how we store and recall sounds, feelings, images, and tastes.

How the Mind Stores Information

Imagine for a moment that at the age of ten years old you started wearing an HD (High Definition) camera with a built-in microphone. Now imagine that you recorded everything twenty-four hours a day, seven days a week for the rest of your life. By the time you reached the age of 40, you would have 262,800 hours of video footage. This would require around 821 pieces of 4TB (4 Terabyte) computer hard drives, which would cost over $100 each at the current prices.

The total storage of these hard drives would be 3,284 TB. That is 3,284,000,000,000,000 Bytes of information. If Micro SD memory capacity continues to grow, at the same rate it has been so far, then all that 30 years of video footage would fit on one memory device by the year 2033. (Detailed facts are at the end of this chapter for those interested in the gory technical details.)

The healthy human eye has an impressive resolution (resolution means detail). Those with 20/20 vision can see more details than HD video has, but HD quality recording would give us a good idea about what was going on at any given time during those 30 years.

Now picture 821 hard drives and notice that if you wanted to be able to find anything you should have these all connected to a good computer. Then you would need some type of sorting program. At least the exact time and date is required along with the 30-year video footage for each frame (individual picture of the video). This would only be useful for finding things you can locate by date. Birthdays, weddings, Christmases, and all major events for which you know the date would be relatively easy to find. On the other hand, if you wanted to find something else and you didn't know the date, you would be lost.

Imagine fast-forwarding through 30 years of footage in an attempt to find something. It doesn't sound like a good solution. Our mind finds things much faster than that, and can recall things as far back as necessary – but how?

Let's assume you have a joke in mind that you can't exactly remember, but you really want to. How would we go about making a search program that can find that joke in the 30-year video footage?

First, we would require voice recognition, as well as advanced face recognition, of every image of the 262,000 hours of footage. To go through that many images, every time we wanted to find something, would take a long time even for a super-fast computer. On top of that, if we needed to analyze the footage frame by frame, it's easy to see it would be quite an undertaking for any system. Just going through, all the data on one hard drive alone, can take a few hours. Analyzing it would take much longer. We would have to speed things up somehow if we wanted to be anywhere close to the mind's speed. By analyzing the recording as it comes in, in real time we can speed things up. We could implement speech and face recognition as the recording comes in. The results of that would have to be stored also on an extra hard drive. This extra drive could store the recognized words, names, and faces. For each word or face that was recognized we would have a growing collection of references to be stored. Each of these references would have a time and date, to indicate when and what was recognized in the enormous footage.

For instance, the word "you" would probably occur very frequently. Possibly 100,000 times or more which means we would have that many time references as to indicate when the word "you" appeared. Less common words would of course have fewer entries. Faces with various features could also be categorized according to hair color, mustache, beard, large, small, etc. These in turn would allow us to search for events by content. A program could use these references of already recognized items. The required search methods would be relatively simple to implement.

Spotting Fakers

For example, if you only remembered that the joke went something like "A wolf, a sheep, and a snake went into a bar …" but didn't remember the rest of the joke then you could start your search program and give it the words wolf, sheep, and snake. Say you also remembered that whoever told the joke, had red hair, but you didn't remember her name. You could feed the search program one more criteria: "A face with red hair". You would probably specify one more thing, which is the duration. For example, that could be five minutes. That is a "face with red hair", and the words "wolf", "sheep", and "snake" all have to appear in the footage within five minute duration. This would most likely narrow the results down to only a handful results to look through. Moreover, why would you be looking for the joke in the first place? Because you remember how much it cracked you up. If the speech recognition software is programmed to detect laughter, then this is one more solid clue you could feed the search program.

Face tracking and red eye removal, are common features in many cameras today, therefore recognizing faces with a certain hair color would be a small step from there to implement.

Speech recognition exists already also. Therefore, this project would be realistic indeed. Given that, someone would buy 821 hard drives, a super computer, and invest in years of software engineering just to imitate a simple ability of the mind. Not to mention wearing a camera 24/7.

Recognized words and faces require much less space to store than video. Therefore, searching on a computer by these contents would be of magnitudes faster than searching through the entire 30-year footage for each search. With proper equipment, the search for the footage that contained the joke is possible in seconds.

Interestingly enough, if you play enough trying to recall old events, you will likely realize that the mind seems to work in a similar fashion. Our mind mainly searches by contents.

Storing and recalling of our mind seems to work similarly to the previous example. The mind also seems to sort experiences as we receive them. Once the recording is sorted, processed, and accepted the mind doesn't seem to look at the result of the sorting again. This is why we don't see false-information unless we are specifically looking for it. Our mind will only find the false-information, when we specifically direct our attention to re-inspect the recordings. The mind seems to have an advanced sorting mechanism in place, which we don't fully understand but with enough effort could mimic certain aspects of it.

The learning by rote phenomena proves that the mind is capable of just storing stuff, without understanding all the contents. When we don't have our attention on something, then our mind seems to analyze less, and store less. We can turn off our cognitive sorting function, just simply by not paying attention. This suggests that attention is the "analyzer" of our mind in reality. What we pay attention to is what the mind analyzes. Normally our attention is in present time analyzing what we perceive but we can direct our attention at will to past events.

This explains why some students do better than others. It is a much better explanation than the conventional assumptions focusing on "smartness" and IQ. Logic would suggest that no one should ever be forced to do something they don't want to. Lack of interest seems to turn off (or seriously impair) the higher layers of our cognitive abilities, where we really connect the dots of the reality we are observing. This is why outstanding teachers inspire instead of teaching. Once we get someone's natural interest up about something through inspiration, all we have to do is stand back and supply them with information and occasional help.

Learning by rote is inferior to conceptual knowledge. Conceptual knowledge is the opposite of learning by rote. It is the knowledge (or subjective reality) that we get, when we recognize and identify, as much as possible from what we are experiencing. This only seems to happen if we're interested. The more interested we are, the more conceptual our knowledge will become. Such deep knowledge is superior to learning by rote. If we do our best to recognize and make sense of everything we can and connect all the dots, then we build conceptual knowledge. On this path, of course research is essential. When we try to connect some dots and it isn't happening, we have missing or false-information. The more dots we connect between the things we observe the more we know.

Every school for every profession has to start with the basics, and building on those basics, the students can move on to more and more advanced knowledge and technology. If the students miss the basics, then they will not be able to understand or perform the later required tasks to the degree they have missed the basics.

Here is a summary of what we know about the mind's storage so far: Our mind is capable of storing large amounts of information. It can sort and analyze the information it receives, and decide how and what to store. It can also recall the relevant information in a split second, so efficiently that it would take a super computer, and a few super programmers, just to mimic even the most basic of its abilities.

Our interest greatly influences our final abilities. Thus, the mind doesn't seem to sort and store everything equally. Only what is relevant and of interest will be analyzed and stored in a useful way. We can force our mind to store raw information, and later recall it, but this will be more like a video/audio recording, and won't be available as useful coherent knowledge.

Gory technical details:

Memory devices and computers are becoming part of our everyday life, so it is worth knowing a little about them. Think of a byte as storage or holding place that can only hold one character (letter). How many bytes are required to store a book then? A 300 page standard paperback has around 1500-1800 characters on each page. Out of 1500-1800, let's take the larger number, to be sure that the book will fit. Better yet, let's round it up to 2000 characters per page just to be safe. The total number of characters (letters) in the average 300 page paperback would be 2000 times 300, which results to 600,000. That's not a small number. This means that storing the average paperback requires 600,000 bytes of storage capacity.

With current memory capacities, $10 can buy at least an 8GB Micro SD memory card. GB stands for Gigabyte, which equals

1,000,000,000 bytes of storage. (Enough to store 1,000,000,000 characters/letters) Micro SD memory cards are the tiny ones now popular in cell phones, as an optional plug-in memory. 8 GB storage capacity works out to a whopping 8,000,000,000 bytes. If we divide 8,000,000,000 by 600,000 then we get 13,333. (Plus change) This means, we could fit about 13,333 average paperback books on that tiny card. 256GB Micro SD card was announced in September of 2012 at the nosebleed price of $899. (These initial high prices radically decrease over time)

Two hundred fifty GB is 256,000,000,000 bytes. That would hold 426,666 paperbacks! Quite a library on something so small, that it's almost too difficult to handle.

Micro SD memory devices have been doubling in capacity about every 18 months. This is an exponential growth. Doubling every 1.5 years means, that in 21 years the capacity would double 14 times. That's not to be confused with multiplying by 14. What we need instead is the 14th power of two, which equals 16,384. Therefore, in 21 years we can expect single memory devices to reach a capacity that is 16,384 times that of the current ones. So how much is 256 GB times 16,384? It is 4,194,304,000,000,000. According to convention, it might be rounded down to "only" 4,000,000,000,000,000. This can happen as soon as 2033. This will be enough for at least 30 years of HD quality footage, all in one tiny device. Will it still be called Micro SD? Most likely not. To access this amount of data, in a timely manner, will require faster communication.

The large number prefix called "Peta" will be required for these. Peta is 1,000,000 times more than Giga. Therefore, in the year 2033, 4 PB (4 Petabyte) memory devices may hit the market.

Do you think science is going crazy with these numbers?

Then consider this:

Imagine a tiny piece of copper cube with edges of only 1/16th of an inch. That tiny cube will have about 338,000,000,000,000,000,000 atoms in it. Scientists are already devising plans on how they will store one byte or more information in a single atom. Considering this, the potential for the future is that the prefix "Exa" will be required somewhere around the year 2045. (Exa is 1000,000,000 times Giga)

Of course, this is all just assumption. We already know better than to accept assumptions as "known" right?

For the technically inclined, here is a quick breakdown of the human eye. According to the University of Pennsylvania, the human retina has a bandwidth around 1.1MB per second. HD video bandwidth is around 2.5-5MB per second, depending on the quality chosen. (Bandwidth is a description of the amount of data per second. A 1MB bandwidth means 1,000,000 Bytes every second, which yields 60,000,000 Bytes per minute). The 821 hard drives required for the 30-year HD footage was calculated assuming 3.46MB video bandwidth.

The proposed recording bandwidth (3.46MB) seems adequate for two human eyes. Since two eyes would require 2.2MB per second bandwidth. Even the most conservative estimates state that the human brain receives over 1MB per second of signals from the *body-sensors*. Human eyes differ from standard cameras greatly. Regarding the human eye there is a spot called the gazing point. This is the spot we are looking at every time. It is the exact center point of our vision, respective to the eyeball. (Not respective to the head or body) The gazing point is where the resolution of our eye is highest. The resolution of our eye is not so good at the peripherals. This may not be easy to notice because we are always staring right at what interests us.

The simple experiment on the next page will demonstrate this.

Hold this book about eight to ten inches in front of your face. Notice the two periods on the left side and one period on the right:

..   .

Here is the experiment with the periods above: While staring at the period that is all the way to the left, without moving your eye at all, notice that you can make out the other period close to it with good details. Now look at the period on the right side of the page. Without moving your eye at all, try to make out the periods on the other side. You may have to practice not moving your eye. Stare at the period on the right, and look with your peripheral vision. See how much of the other periods (the ones on the left) you can make out without moving your eye. Don't look at the double periods look straight at the single period on the right!

168

Spotting Fakers

What you should see is that you can make out the periods on the left but with much less detail. (While staring at the one on the right) Even if you have very good vision, the periods will appear as one or almost as one. This is normal. You can see them but not in enough detail to differentiate as well as when you look straight at them.

This experiment demonstrates that the human eye isn't like a video camera at all. A video camera records the image as a bunch of little pixels. In case of a 1080p HD movie, that's about two million pixels for each image (aka frame). The pixels are uniformly sized, and arranged into a rectangular shape. Our eye would only be similar to a circular shaped camera, instead of a rectangular one. The eye has higher resolution sensors, in the center. As we go from the center of the eye towards the peripherals (that is the edges of our vision), the resolution of the eye gets lower. This is why we can't make out the dots as well, when we are looking at them with our peripheral vision. Naturally, we don't notice this, since we move our eyes constantly as necessary in order to see what we have our attention on. Our eyes move about constantly scanning the edges, patterns, and shapes of objects and people, we are looking at. If we wanted to have a recording similar to the eye, we would require similarly built cameras that mimic all aspects of the human eye. These would have to look at what our eyes are looking at, all the time. This would require gaze-point tracking technology, which senses what we are looking at. (Existing technology by the way) The resolution of the cameras would have to drop off towards the peripherals, just as our eye's resolution does. Talking about this science isn't a far into the future Sci-Fi subject. Scientists have already made breakthroughs, with chips implanted into human eyes, to help those who were born blind or lost their vision. It could be theorized that the 3.46MB per second bandwidth might be enough for human-like recording, or at least it's comparable in magnitude, to the information the body senses at any given time. All bandwidth numbers related to the human eye are merely estimates for the sole purpose of giving context to the theories.

The whole purpose of this breakdown is to show just how much information our mind processes nonstop, and point out how difficult it would be for technology to mimic what we can do. The examples shouldn't be considered as an attempt to explain the physical form of the mind. Our focus is on the functionality of the mind.

(Technical note: All bandwidth values converted to "MB per second" for simplicity)

Skill and Recall

The subject of how we remember and recall past experiences is an interesting and important one. Sometimes it takes a while to remember something. We may find ourselves digging around and still come up with nothing. Then look a little more or just walk down the street, and all of a sudden, "click", there it is. Whatever experience (or memory) we are looking for, once located, it comes in as if it happened yesterday, no matter how long ago it was. (As described in the *Timeless Mind* chapter)

People can recall experiences that happened to them 20, 40, or 60 years ago and are able to describe them in vivid detail. This brings up the question: Do memories really fade with time? How is it that some people can recall vivid details on certain things but not others?

Here are some possible answers:

It's possible that the mind selectively "throws out" certain memories, perhaps after a while. It is also possible that the memories we can't recall are still there we just don't know how to find them in the vast jungle of our *mental-copies*.

The main purpose of the previous chapter is to show the incredibly huge amount of information our mind has to deal with on daily basis, as well as the complexity of having to find something in such a large collection. Is it possible that we carry more information than we can recall?

We can make simple everyday observations for ourselves in order to draw a conclusion. If we look around, we can observe that those skills that we quit practicing for a long time tend to fade away. An obvious one to observe is language.

171

Miklos Zoltan

Language is a skill. It has all the characteristics of a skill. To be good at it we learn and practice it for years. After immigrating people are usually required to learn a new language. If we study those who stopped using their native language, we will find that it is fading. Not entirely of course, but as time progresses certain words don't pop in as easily as they used to. Then as time passes without practice, after a few decades it may fade so much that speaking it becomes difficult or even uncomfortable. At that point, it may seem that the native language is being forgotten. Then a visit to the old country after 20 years would make the native language required again daily. In such case, knowledge seems to come back unexpectedly. After a week or two of intensive communication, the language skill recovers almost to its original state. Now all of a sudden it seems like the language was not forgotten after all. This also holds true with other skills. Any skill that is not being practiced is subject to some fading.

We should be objective and careful when studying such phenomenon. For example, people who live their lives with a new language may still be practicing the native language while talking to their relatives over the phone. This would keep their native tongue alive. Only those who didn't use their native tongue for over a decade (at all or hardly ever), should be observed to see if the native language truly fades.

Some may not readily admit that they have lost command of their native tongue for various reasons. Then again, some may exaggerate how much they lost. Physical ills related to memories may affect language also. A few decades can bring much change in the native language. New words to keep up with technology and new slang alone can make speaking the native language quite challenging after 20 years. To make accurate deductions, we should rule all such cases out.

Great attention to details should always be a priority when collecting information for our conclusions. When we study a particular subject in search of a pattern, we should rule out anything that is irrelevant. Patterns are essentially predictable similarities on a given subject. Therefore, items that don't strictly belong to that subject should be ruled out.

172

Spotting Fakers

For example, let's say we want to compare apples to apples, from a large basket of assorted fruits. First, we should sort and pick out all the apples. Then we may find that there are apples of different colors. After picking the color green, for instance, we could sort through all the apples and pick out only the green ones. Then we can truly compare apples to apples, which in this case would be comparing green apples to green apples. Then we can attempt to find some similarity between them, other than the fact that they are all green. For instance, we may discover that green apples never have more than six seeds. (This is an analogy and not a thoroughly researched fact)

The idea is to define a certain group of something. Then find something else that is always true for that group. That something else should be true for every single element of the group. No exceptions. If we found even a single green apple that had seven seeds, the six seed rule would immediately be false. It would have to change to "green apples never have more than seven seeds".

Noticing differences is just as important as finding similarities when we look for patterns. Our minds mainly look for patterns.

Studies and statistics are based all too often on similarities without enough emphasis on dissimilarities. The ability to notice dissimilarities is also as a skill. Having that skill perfected is useful for everyday life, as it is a way to avoid generalizations, stereotyping and false assumptions.

Looking at our abilities from the viewpoint of skills is very revealing. With this view, abilities previously thought to be genetic, are visible for what they really are: skills. The conventional idea that ability and potential are genetically determined doesn't seem obvious. Genes and physical structure may have an effect on our mental abilities in a global way but most of what we do can be seen as skills and ability we adapted.

Most of what we can't do can also be explained as a lack of skill as opposed to lack of physical structure or genes. The best way to show the potential difference is by another analogy.

Instead of thinking of the mind as a "tool" that is pre made for certain tasks, it seems more feasible to think of it as a general "tool" for thinking and solving problems (adapting). This "tool" is unlike computers. Computers can only do what we program them to do. The mind seems to be able to make its own "programs". It can make a program for playing the piano, math, welding, swimming, martial arts, dancing, or whatever we desire. The mind creates the "program" as we progress with the activity. The skills we acquire do not depend on preexisting conditions (genes). We acquire skills as a result of demand and opportunities of life itself, combined with our personal interest. Whatever it is we have our attention on, our mind starts creating the necessary "program" for.

Adaptation is what the mind can do, and it does an excellent job at it. Can it be bogged down? Yes, of course. False-information is the mind's number one enemy. Fortunately, our mind is capable of making a program even for finding and fixing false-information. In terms of computer analogy, if we had computers that could adapt themselves to solve any problem, by writing their own program to handle the required task, then we would have something *similar* to a mind.

Computer analogies are used because it is the closest technology to a mind, for the time being. Computer science is based on well-defined and widely understood facts and is *not* a gray area of knowledge. Thus, computer technology is the best we have, as far as analogies for the mind, since computers are the only devices that we understand that can implement logic, and mimic intelligence.

The study on skills suggests that forgetting doesn't mean things have been erased from our mind. Skills rely on our memories and experiences. Memories, experiences, observations, mental images, opinions, thoughts, skills, feelings, etc., rely on *mental-copies*. Whatever we do that requires learning, which is essentially everything we do, requires ample amount of *mental-copies*. We need these to be present and available for recall in our mind for everything we do. Additionally, these have to be logically linked to one another (analyzed) by our mind, which only happens when we pay attention.

Spotting Fakers

When these *mental-copies* aren't being used, our mind seems to move them to a back storage. This is similar to a secretary moving files from the office to the file storage room. Once things aren't frequently used, the mind seems to degrade their importance so to speak, and withdraw attention. If that unused knowledge becomes necessary again, the mind requires time to bring that knowledge up front again. The equivalent of this would be the secretary having to go into the file storage room and look up the folder of an old case. This could be thought of as a re-build or re-link process where the mind throws its "hooks" back into the *mental-copies* required for that skill. Thus, the knowledge becomes usable again. This process is much faster than what it took to learn the skill. This would suggest that sometimes we may seem to have forgotten something but it's still there. We just have to wait until our mind finds its way to that particular section of our immense collection of *mental-copies* that is required. The key is not to be dismayed by the information not being available right away. We shouldn't jump to conclusion too quickly and think the memory is gone. Instead, we should keep our attention on it and insist that it is necessary. This is automatic if we start practicing the seemingly forgotten skill frequently.

Language tells us a great deal about the internal workings of the mind in more ways than one.

The native tongue is what we pick up first in our life, as verbal and written communication. Even if we don't use it for a long time, we can eventually recall the meaning of words, sounds, concepts and everything else across decades. All of the above would suggest that the mind doesn't seem to have an issue with recalling over time. How long ago something was, doesn't appear to matter much. Once we get a grip of a memory, or some element of the memory, then the entire experience gets recalled with potentially all of its elements across time. In case of the native language, since it is a large collection of information, it seems that large chunks of that collection gets brought back into focus, when we reengage in that seemingly forgotten skill. (Secretary moving hordes of old files back into the front office would be the equivalent)

It seems logical, that when we get in trouble with recalling, the problem is not related to how long ago we acquired the *mental-copies*. The problem seems directly related to lack of relevance. Finding the proper link, to memories, and experiences, is what takes time. When *mental-copies* aren't used, the mind ranks them as less relevant in relation to everyday life. The things that we use frequently remain easily accessible.

No two people are exactly alike. Therefore, we don't observe the universe the same way. One can be a music genius, with perfect hearing but poor visual recall. Another can recall vivid images of old school mates from decades ago but have no musical inclination. Then again, some might have it all. It all depends on their adapted interest and experience as opposed to genes.

It is logical that those things we all do, and none of us is an exception, stem from underlying physical or otherwise fixed causes. Recalling does not appear to be one of these.

Some of us are efficient at recalling at will. Others may have a hard time recalling old events. When they finally do recall, the memory comes in as fast as something that happened the day before.

The important part is not to confuse the time it takes to locate a memory, with the time it takes to access (use) it.

Recall is a two-step process. Step one is the action of locating the required *mental-copies*, which contain the memories or experiences we are looking for. Step two is to access these *mental-copies*. Locating is the process of finding the proper link, association, path, or reference. The action of locating something has the characteristics of a skill. The more we practice it the easier it gets. The action of accessing *mental-copies* seems near instantaneous, (a fraction of a second) and thus has the characteristic of a physical property of the mind.

The time it takes to access *mental-copies* may slightly vary from person to person, but compared to the process of locating it is negligible. *Native-thoughts*, (the unvoiced ones) which are below our awareness, use our *mental-copies* all the time. These thoughts are fast. They only take a fraction of a second. Our mind can have 5-10 (or even more) of these thoughts in under a second. If these all rely on using our *mental-copies*, then this suggests, that accessing *mental-copies* is similarly fast, requiring only a fraction of a second. In which case the delay of recalling is likely the result of us looking for something that is not used enough, which is to say it has not been accessed enough, and thus the mind considers it irrelevant. What takes time for the mind then, is to locate the information, which has been "stamped" irrelevant, but is now required again.

Example: Someone tells a joke. The listener almost falls off the chair from laughter, and says, "I have to remember this one". Two days later he wants to tell the joke to a friend, but can't remember it. When he heard the joke, he was laughing so hard that he couldn't make a real effort to remember it. "I have to remember this one" became "It would be nice to remember this one". If this person is not well practiced in the skill of recalling then he will likely think: "I can never remember jokes." Such considerations can work against prompt recall at will. He might just call that friend, who originally told the joke, and have him repeat it, perhaps with pen and paper ready this time. How did the person who told the joke remember it then? Probably liked it enough to memorize it, or perhaps, that person was already good at recalling things in general.

To be a professional at any skill, one should practice it for years. If recalling is a skill then those who practiced it regularly will seem to others as having a "better memory".

Some people get into the habit of recalling at an early age, because of their specific life experiences, and do it frequently throughout their life. After a few years they will be very good at it, and have years of a head start on others. Of course, they will seem like they have a "better mind".

If you ask people with such ability, how they can recall so easily, they will probably tell you that they don't know how. If we question a little, then we will discover that they started practicing recalling things early on, and gradually got better. They just never realized that, since it's so obvious to them. Understandably, then by the time they reach the age of 20, they are pros at the skill of recall. To make matters more interesting people can have excellent recall in certain subjects while not being able to remember other subjects. One can be excellent at reciting (recalling) jokes at will, while having a hard time remembering the names of people they meet. Personal interest is likely the underlying cause to such differences. Interest can define our paths early on. Chances are if we asked such person, we would find that they are indeed interested in jokes but not so much in names for instance. We might even get answers like "Who cares about the name. I can remember the face and personality. That's what matters." Such reveals that the faces and actions of people are what's important to this person, not so much the names. In that case, it wouldn't be surprising that he can't remember names. We can improve our ability of recall, but to be excellent might take a dedication that is comparable to learning a second language.

If we were to ask anyone, how they learned their native language, we would likely get shrugs and wrinkled foreheads. Then the answer is usually: "Hm ... I don't know; I just it".

So, how did <u>you</u> learn your native language?

When we learn something and without planning pick up the skill, we have no idea how we did it. The reality is that our mind perceived some patterns, which was frequent enough for it to start adapting to it. All these "sounds and pointing", and "sounds and action", we experienced as infants contained patterns. We figured these out because this is what our mind does – adapt. If this is so then those who are good at recalling, might have started picking up that skill at an early age. This could be for the simple reason of frequently seeing others being able to recall or getting the idea that that it is useful. If we ask them about how they got that skill, we'll be likely to get shrugs *again*, for the same reason. They do not know how they did it, they just did. It was an interesting "pattern to be cracked" for their mind.

Spotting Fakers

How much of what we see is recorded as *mental-copies*, and how it is sorted varies from person to person. Since our individual interests can affect how we process information, it's logical that recording is also affected by how interested we are in what we experience. When we're disinterested, even our head and eye movement slows down, and hence we don't take in as much of the reality around us, as we could if we were interested.

Consider sitting in a movie theatre. If the movie is interesting you'll be glued to the screen and absorb all that is happening. If you are into it, you'll feel like you are in the set. Even after the movie, you may play around with the scenario and characters in your mind and wonder what it would be like to experience what you saw. If you are not into the movie, you may end up texting through the whole flick. Life is like that. We receive the images of our life through our *body-sensors* no matter what, but we can turn our attention from it at will.

This is why it can be life altering to be able to gain control of our own interest.

Consider this: The movie playing might not be your favorite genre but since you are sitting there already why not see what you can learn from it that might be useful later. This applies to life experiences also. The idea is to make the best of any experience. We can learn new things at the most unexpected places.

If something is forgotten we don't know if it is really erased from our mind or not. What we do know is that practice and interest combined can greatly improve our ability to recall many things, which seem forgotten. The only way to prove this to ourselves is to find a subject we have a hard time recalling, and see if we can improve our ability. Practicing recalling has to be done daily while keeping in mind what interests us. To be better at recalling names? Jokes? Faces? Science articles? Last minute grocery lists? There are plenty of free exercises available on the Internet, which work very well. There are many documented cases of people, who increased their ability of recall. This proves that recall is a skill, which can be improved with dedication.

Linking

It is a fact that our mind records images, sounds, smells, tastes, motion, position, as well as internal and external sensations throughout the body. We can observe this to be true and logical. We can recall past events, but we can only recall our own experiences. We can't recall someone else's *metal-copies*. Even if we were at the same place and same time as someone else, we are recalling our copies of the experience.

To the mind, everything is an experience, even when we are sitting still, reflecting on past events, meditating, or sleeping. How do we know that? Because we can recall these activities also, which means they became part of our *mental-copy* collection. The recording is unique to each of us. How much of what we experience is recorded, depends on our interest and curiosity at any given time.

Everyone is recording his or her own version of reality. We aren't concerned with where this information is stored. It is self-evident that it is stored somewhere.

It is also a fact that we can sometimes make mistakes. Following the logic presented so far, it can also be considered a fact that poor information (containing any falseness), adversely affects our judgment.

The final step the mind has to accomplish while processing information is to find patterns in what it receives. Information that is already sorted and accepted, isn't rearranged, or analyzed unless our attention is directed explicitly to it.

Our mind is capable of recording our experiences, and make predictions based on these recordings. Whether we are aware of it, or not, whatever we experience causes our mind to access the recordings that are most relevant to what is going on in present time. The content of these recordings are stored in time sequence, which we know because if we recall two consecutive events, we can tell which was first. Additionally, our mind categorizes experiences by similarities also. This is an associative linking process. While we aren't concerned with all the structural details of how our mind does that, there are certain patterns that can be observed about how this linking works.

The mind links things in a sequence. This is not to say all life *experiences* are linked together in one lifelong string of events. Instead, experiences, people, objects, concepts, etc. are connected into little bunches of recording sequences and arranged based on relevance and similarities. The mind does this linking in a way that the original sequence of events remains intact also. The vast majority of these recordings appear to be split second in duration, but this is just a side note so we can have some idea.

What makes linking important is priority. The concepts that we have on a given subject, don't receive the same priority.

The earliest concept of any given subject takes priority.

This is what our mind will access at all times, when that concept is required. When the mind encounters a new experience (or concept) that is similar to something we already have, it will link (or associate) the new event to the old one. This is similar to building a structure. The foundation goes down first then everything else is built upon it. For this reason, the mind always relies on the earliest concept. If the early concept contains false-information, it will "infect" all connecting future observations. The only way to fix this is to find the false-information in the earliest concept and spot and recognize what makes it false. What we are doing in such case is realizing how and why the false-information is not in alignment with reality. Once we spot the deviation from reality, it's handled. The mind will do the rest to correct it. The best way to show this is with another example.

There is a common misconception related to astronauts. If we ask people the simple question, "Why do astronauts float on the

International Space Station?" a surprisingly high percentage will respond with the following misconception: "They float because they are far away from Earth and hence gravity is not affecting them".

In reality, they float because the space station is traveling at a speed of 17,500 miles per hour zipping around the planet. That speed causes the space station to want to go straight and leave the orbit. Earth's gravity however, doesn't allow it to do that because about 90% of Earth's gravity is still affecting the space station. The two forces balance each other out which is why the astronauts appear to float freely.

Let's see how people get the misconception.

The space station is around 260 miles above the surface, which is extremely high up for us, especially when we consider that the highest peak of Mount Everest is "only" about 5.5 miles. When we compare 260 miles to the size of the Earth then we realize why gravity is still affecting the space station. The diameter of the Earth is 7926.41 miles. Compared to that, 260 miles is just "barely" off the ground.

During childhood, we watched TV footage of some rocket launch (usually one of the Apollo launches) until the rocket was so far, that it was no longer visible on the TV screen. Then later, we saw footage of people in "outer space", floating around playing with floating water and such.

The appearance of the immense distance in which the rocket was heading, can give the illusion that they went way out there, far-far away from our planet. In addition, expressions like "outer space" can easily support a misconception of distance. The concept we tend to create is something like "they went far away from the planet, so they don't experience gravity."

The even more interesting part is that after clearing up the misconception, the majority of people remembered learning about the underlying scientific reasons. Despite of that they were still thinking with the misconception whenever they saw or thought of an astronaut in space. In short the misconception was "Floating = no gravity".

This is an obvious case of a *false-known* caused by missing information. The problem is that the early concept doesn't necessarily get corrected by the science that is learned later, unless we spot the earlier false assumptions and the false-information that was causing it.

It doesn't matter how much more we learn on the subject if we don't spot the false-information. Floating in space and gravity in general, will cause us to associate to the same *false-known* time after time.

Not everyone has the same misconception. Those who gave the correct answer had someone available to explain the underlying physics and words used until the false-information was handled. Words can be misleading at times, especially generalizing terms.

This example demonstrates that our mind has a tendency to use the earlier concept, whether true or not. This way later knowledge can be infected and cause us to make mistakes or get bogged down with the subject.

Additional information will not automatically fix the misconception. Research also indicates that this is more than a tendency, it's a rule.

Our mind uses the earliest accepted knowledge, on any given subject. Whether the knowledge is false or not, doesn't matter.

If the knowledge is false, it can only be corrected by finding the underlying false-information. As that happens, the mind will continue to use the same mental images, but it will rearrange the necessary logical links. This way the *false-known* (misconception) becomes a known. In analogy of a building, this is equivalent to fixing the cracked foundation. Except in a building, after fixing the foundation, the cracked walls or framing may also need fixing. In our mind, the repair work is automatic. We may even sense this as it is common to feel a bit "spaced out" after making such cognition. Since the mind's linking is automatic, it is possible for such cognition to create sort of a chain reaction of other cognitions.

Much or all of that can be subconscious and never voiced but we can feel it. The linking process of the mind, can interlink just about anything in which it finds similarity. The rule is that when the mind is desperate and missing or omitted information is present, then any similarity will do. This is the foundation of assumptions. For any given concept there is always a first time, when the concept was conceived. This concept is then built upon like a foundation. Everything else related to that concept will link back to this original one. These first concepts of any given subject will be referred to as *base-concept*.

A *base-concept*, which contains false-information is a *false-known* and causes all knowledge consecutively connected to it, to be false. The reason for this is the linking nature of the mind. Linking is the process of the mind connecting (bunching together) similar events and experiences into a sequence. The mind has many of these sequences. There is a sequence for every subject we ever conceived. When we experience something, our mind finds in a split second the most similar sequence of concepts. Then it recalls (accesses) the *base-concept* of that sequence. In the example of the astronauts, the earliest *base-concept* having to do with astronauts and gravity was that "floating means away from Earth". Therefore, it doesn't matter why the astronauts were really floating or what caused the "antigravity". What matters is what our concept says. (Floating = away from Earth) This sequential dependency reveals how a single false-information can affect so much of our judgment and thinking. This is how fakers are affected by their rule #1 (others do *not* matter) throughout their life every time they deal with others. On the concept of others, the *base-concept* they recall automatically (without being aware) is that "others don't matter".

Native Thoughts

Native-thoughts are not so obvious to spot for a number of reasons. These are at work whether we are aware of them or not. We can learn just about any skill, without knowing the inner workings of the mind. The only things our mind doesn't start learning about are the ones that we don't pay attention to and the ones we don't experience.

The study of how the mind works isn't considered to be a *must-know* subject, probably because we can learn most things without such knowledge. Many of us can achieve what we desire without a manual to the mind. The problem is that this is only true for our own mind; we can't see into someone else's. This is why it is useful to know a little about the human mind in general, as we can use that to understand others as well as ourselves. This allows us to see a larger picture.

The adaptation process of the mind works on its own creating the necessary thoughts (concepts and links) for any activity.

During the course of learning, our mind will have tons of *native thoughts*. *Native-thoughts* aren't apparent as we don't voice them. Which means we don't slow down enough to perceive them. If we only had ten every second, it means we have about 36,000 *native-thoughts* every waking hour. They are the means of the mind's internal logic at work. Some refer to them as subconscious thoughts. This isn't the best expression though since it may imply that we *can't* be aware of them. That notion has the potential to cause us to think that these thoughts are not to be trusted, since we aren't aware of them. The truth is that whether we are aware of their existence, or not, these are the thoughts that solve problems.

187

Contrary to common belief, voiced thoughts are not necessary for us to solve problems. Voiced thoughts are only necessary when we are communicating with others. The mind not only does perfectly well without voiced thoughts; but it is much faster and more effective when we don't voice.

Think of voicing thoughts in our head as an extra workload for the mind. *Native-thoughts* are not inferior to voiced thoughts by any means. They are superior. They are cleaner and faster.

Native and voiced thoughts are equally sensitive to false-information.

Therefore, false-information will equally cause trouble for both types of thoughts. Just because we voice something in our head while thinking about a problem, doesn't make our thinking process any more, or any less susceptible to false-information. Voicing only slows things down.

The sign of *native-thoughts* being at work is a quiet mind, yet there is a good feeling of contentment and progress.

Voicing thoughts while thinking is an unnecessary habit, which develops when we don't trust our mind. What happens is that due to false-information, we are prone to make mistakes. Once we realize the mistake, we start accusing the mind of being at fault. Then we force the mind into voicing the things in our mind with force, to give those thoughts emphasis. In turn, this just ends up making things even worse, as the mind now has to slow itself down also by voicing every thought.

Since voiced thoughts are the ones we can perceive, they can lead us to the misconception that this is how we think. It isn't. Native thoughts are quiet and we should be able to trust them.

Attention paths

Since we tend to base our views on our subjective reality, it is natural that we pay the most attention to these:

- Things others do to us.
- Things we do to others.

These are the two main and most obvious paths of our attention.

There are two less obvious paths:

- Things we do to ourselves.
- Things others do to others.

Because of influence, we can do all sorts of things to ourselves. As mentioned before in examples we can end up falsely judging or accusing ourselves when someone with hidden goals reveals their judgmental opinion of us. This can be demoralizing. Therefore, when we receive such indications, judgments, or evaluations about our actions, personality, or integrity it is best not to accept it without proper verification. Understanding the source of such negative remark is important in order to keep our mental freedom.

When others do things to others, it is easy to dismiss what happened thinking it's none of our business. If we wish to watch out for each other, it makes sense that we should pay attention to and spot, when others are treated unfairly. If it looks unfair, it probably is and we should look into the details whenever we can. Even if we aren't the one directly responsible for the persons involved, just simply acknowledging the unfairness can help the victim stand up for themselves. When we tell others about such outpoints they know they are not alone. When we ignore the unfair treatment of others, we may be permitting fakers to thrive. Just simply speaking up about unfairness we observe, we can take away the power of a faker.

Fakers promote the idea that "it is none of our business". This way when they take advantage of others we give them less trouble. They have been promoting this idea for so long that they managed to embed it into our collective consciousness. It is just another one of their illusions. We should watch out for others to the exact same degree, we would like others to watch out for us. This can be derived from the golden rules:

"One should treat others as one would like others to treat oneself."

"One should not treat others in ways that one would not like to be treated."

Persistent Problems

As mentioned before, persistent problems are frequently (or mostly) due to false-information. When we know the solution to a problem, it is not considered a persistent one. Having to work for a goal should not be confused with a persistent problem. A goal may take decades to achieve, but as long as we see the path to the solution, it is a *work in progress* and *not* a persistent problem. When we don't see the path to the solution no matter how much we try then we have a persistent problem.

What makes a problem persistent is that we are *not* looking at the real source of the problem. It has nothing to do with our physical abilities. We are all able to solve problems and connect dots we just need to look at them differently. This chapter explains how to do that.

When we are looking for the source of a persistent problem we'll be asking ourselves various questions in order to locate the source. These questions will help direct our attention to where we should look. You can and should tailor the questions to your preference and specific problem. In order to do that successfully it is crucial to understand the reasons and follow the rules. Please don't follow the guides because you believe they are correct. Only follow them when you see them to be correct.

Rule #1 of persistent problems:
Avoid the word "why" in your questions when looking for the source of a persistent problem.

But why?

The reason for that is logical. The word "why"' directs our attention to already existing knowledge we have. Which means it is causing us to keep looking at the results of our already existing and accepted logical predictions. In other words, it directs our attention to what we already know. We can only answer a "why" question if we already know. Since false-information can cause us to have *false-known* it makes sense that even one of these can prevent us from answering the question and it can make us go around in circles. This is because we are only looking at what we know, which is all stuff we already accepted and "stamped" as trustable. The falseness of a *false-known* is not visible to us, which is why we can't solve the problem this way. When we can't answer the "why" then we have a persistent problem. Keep asking "why" will make us go around in circles looking at the same "known" items over and over again without being able to spot the *false-known*. (Loop thinking)

Persistent problems can make us feel trapped, foggy, groggy, tired, or overwhelmed. This can make us feel like we have a splinter in our mind over time because the problem is unsolved. These are obvious signs that can tell us that we have a persistent problem or that we had one in the past that is still lingering around and affecting us.

Understanding the priority of facts, logic, prediction, and known comes handy for persistent problems. (From the chapter *Knowing*) It is the key to understanding the cause of persistent problems. Normally our mind uses what we know. This means that we normally deal with stuff that belongs to the highest layer of our knowledge, the "known" layer. These are considered verified and we don't normally question these. If one or more is false it can prevent us from solving the related problem. Asking "why" keeps our attention at the "known" layer which means we are not going to look at lower layers. Persistent problems are caused by false-information. False-information is at the facts layer. Which means we are just simply looking at the wrong layer when we keep asking "why"? We can direct our attention to lower layers by using objective words like "who", "whom", "what", "which", "when", "where" etc. Asking questions with objective words only, directs our attention to the objective layer of our subjective reality, which is "facts". That's where the false-information will be. That's what prevents us from seeing who or what sidetracked, manipulated, or influenced us. That's where we'll find the source, where the false-information came from.

The way we word questions can make night and day difference.

Our mind deals with the words we hear, and uses our concepts accordingly. We can word the questions to any problem as needed and direct the attention to the first layer.

"Why" isn't the only word that should be avoided. Any question, which is not objective in nature can get us into trouble and keep us on the "known" layer.

Examples to avoid:

"How is that possible?"
"What is the reason?"
"How can it be?"
"Why doesn't it work?"
"Why can't I solve the problem?"
"Why do I have to deal with this?"

Aside from these, we may also have thoughts such as "This is so mysterious" "This is mind bending" "This is impossible" "I don't see how" "I don't get it". Thoughts like these are sure signs of persistent problems also. There are too many combinations of these to list. In addition, words like riddle, secret, and mystery can keep our attention on the "known" layer. These should all be avoided.

The key is to phrase our questions so that they address objective real experiences, interactions, people, and information.

By inspecting the things we find with these questions, we can find the false-information. Keep in mind that there might be more than one false-information at work. Therefore, this is a repetitive cycle. When you find a false-information then inspect it until you spot where it came from. Also spot who or what makes it false. How it deviates from reality. Then it's handled. If the problem persists, then keep looking for more false-information.

Inspecting the contents of the "facts" layer requires curiosity. It shouldn't be based on suspicion or mistrust. These can make us not trust our facts in general. That is not good because we need our facts for knowledge. Suspicion or mistrust can shake the foundation of our knowledge.

When we get bogged down it may seem, that it must be a bunch of false-information causing it. This may make us suspicious of all of our facts. It is important to keep in mind that one false-information can make a whole subject foggy and cause a persistent problem. Usually we have just one or two false-information on any given subject. We should resist the urge to suspect that there is more than one, or that it is something complex. We may have more than one false-information on complex scientific subjects but even then, our correct facts outweigh the false ones. This is why it is not necessary to get over suspicious. Instead, we should look at our facts with genuine curiosity and intention to verify.

Rule #1, Others *do not* Matter

Now that we've covered some fundamentals of the mind and false-information we can look at what drives rule #1. For best understanding of fakers, it is useful to look beneath the rule. Accepting lying as a way of live is the key to why fakers do what they do, whether knowingly or subconsciously. Lying is essentially an intentional alteration of reality. Statements, which are in good alignment with reality, are truths. Truths are unaltered reports or observations of what someone perceived. For an observation to be true for others, it should describe reality well enough. Lies are incorrect reports about reality. A lie is essentially an intentionally altered report about reality. When it is unintentionally false, we call it a mistake or error.

The problem fakers have is that they have picked up a habit (probably early on as a child) of not verifying their facts thoroughly. Therefore, it is questionable whether they are really lying or not. The truth is that some of them do it intentionally while others are just simply too inaccurate in their observations and therefore their reports are faulty. Such faulty reports affect their judgment. When they accept an inaccurate observation, such as "others can't be trusted", "or others lie" it becomes their *false-known*. As a result, when they spread illusions they are not always aware of doing so. Many times what we perceive as lies is just them sharing their own incorrect (altered) observations. One way to look at this, which may simplify what is really going on, is to realize that their subjective reality is inaccurate. It's not paralleling reality enough. If we were to compare their subjective reality to the objective reality, what we would find is endless alterations and differences. They have a bad habit of altering things, as they

195

perceive them. Altering reality is a persistent trait of fakers. When they pretend to be friendly for example, all they are doing is altering their own true intentions. For example, if a faker is planning a rip-off his true intentions are hostile. He alters that intention and displays something different towards the outside world, such as contentment or happiness. In essence, what fakers are doing is altering facts and reality most (if not all) the time. Understanding their actions from this viewpoint can greatly simplify what is really going on. Their problem is simply that they alter reality and therefore they think with an altered version of reality. They don't necessarily alter every subject. The subject of social living and other people is where they alter for sure. Because of their rule #1, they end up altering most of what they observe that has to do with others.

Honest people's reality about others is mostly unaltered. Considering the amount of false and omitted information we deal with daily, it is challenging enough to do a good job of assimilating things unaltered. The last thing we need is to alter it even more. False-information is essentially an altered report or perception of reality. Consequently, the difference between honest people and fakers is as simple as, how altered their subjective realities are on the subject of social living (others).

In reality fakers can a do perceive certain things unaltered. A subject *not* connected to social living may be something that a faker can see unaltered. (Although this is uncommon) Similarly honest people can and do alter reality from time to time, but not nearly as much as fakers. Honest people have a much more accurate reality on the subject of social living due to "other DO matter".

A theoretical threshold exists in relation to altered subjective reality. When a person's subjective reality contains too many alterations (that is too many deviations from objective reality) they might not be able to improve on the given subject until they find enough of the underlying false-information (alteration of reality). In other words having too much false-information on a given subject can prevent us from learning that particular subject. This can easily be fixed by finding false-information. Once we fix enough false-information on the troubled subject, we can study at ease again. When someone is below the threshold it means, that their subjective reality contains so much alteration, that they cannot perceive new information without altering the new information. It doesn't take a whole lot of false-information to have such trouble on a subject. A single false-information is enough to create a significant misconception if it is deep enough. For example, a false *base-concept* is more than enough to give us significant trouble on the related subject. (Chapter *Linking*)

Fakers are below this threshold on the subject of others because of rule #1.

This can be a catch 22 situation. This condition is unfortunate. He may be trapped indefinitely. Being below the threshold isn't what causes fakers to be stuck on this subject. They are stuck because of their crimes against others. They could look for and find their false-information on the subject of others and society. The problem is that the moment they rise above the threshold, they would bounce back down as they realize their crimes. This will force them back below the threshold, unless for some miraculous reason they can forgive themselves. Then they could recover. Start a clean slate with life and others. We can forgive them, which may help, but in the end, they have to come clean with themselves. Without that, they will bounce back down below the threshold and keep altering stuff (reality).

This means that the subject of others (social living) is an exception. Being below the threshold on any other subject does not cause crimes against others. Being below the threshold on the subject of others causes him to commit crimes against others. This is what creates the trap.

It is important not to fall for a faker who fakes to have gotten better. In case of such a claim, we should continue to observe the signs and see for ourselves.

Fakers may be perfectly fine and well above the threshold on other subjects. They could be good at math, astronomy or computer science for example. Anything that is abstract enough and not connected to social life and humanity, they can master. They usually don't get extremely good due to the intersection of these subjects with social existence but isolated cases do exist where a faker becomes really good at a profession while not being able to see the big picture about humanity and coexistence.

Application

Viewing things from the perspective of false-information is a unique tool, which inherently requires us to break events down to elemental components. Looking at the problems of people, societies, and humanity in general from this perspective, opens up a new way to find the solution. Considering that omission is also a form of false-information, suggests that the vast majority (if not all) of our social problems can be explained.

Fakers are a special case of problems caused by spreading false-information, which is the result of their own false views. It's these views that empower their outlook on life.

By allowing fakers to manipulate us, we grant them power. Listening to a faker and buying into the illusion they offer is a form of granting power. The moment we don't spot the illusion it will have the potential to influence us and do something the faker wants us to do. This can sway us from doing the right thing. When we see a faker working on exploiting another, by doing nothing we also grant power to the faker. If we allow them to have power they will use it to get what they think they deserve, which is just about anything they can put their hands on. If we give fakers trust, betrayal is what we will receive in return.

If you are in a position to stop a faker, by not giving the power and trust they seek, you will help all those around you who the faker would have abused with that power. No other retribution is necessary. Spotting fakers doesn't mean turning our world into one big poker face sitting contest, to see who slips up first. We can all go about our business as usual, as we did before we became aware of fakers. When we are ready to spot fakers, what we are doing essentially is paying attention to the actions of people around us.

Spotting fakers means to judge people by their actions, instead of judging them by the words they say and "masks" they put on. When the time comes refusing is all that's required of us.

Shared World

Our world is a shared playground for all of us. No two people can be at the same place at the same time, ever. No two people can be doing the same exact thing, at the same exact time, either.

Sharing, and how to share, is and has been a problem as far back as documented history. Sharing is best when we "play nicely". Most of us get this, but fakers, don't. Before we get into this subject, let's point out the obvious.

Sharing, in and of itself, and having to share, isn't the problem. Those who are unwilling or incapable of sharing fairly are the problem.

Fakers are unwilling to play nice, but pretend to do so. This is a result of their rule #1. Life is like a game. We play it as kids, and we play it as adults too, it's just that the rules, customs, and expectations change when we grow up.

Sharing is such a big issue for us that we had to invent money in an attempt to solve it. Some see money as a convenient solution. Could it be that it was an attempt to solve a huge problem?

We know that people in communities began trading with one another a long time ago. If fakers were involved in these trades, one party would always feel like they didn't receive enough exchange. What type of people would feel like they didn't get enough? Those who feel irrationally entitled. (Sign #6) Since the ability to act is as old as humanity itself, we can safely speculate that fakers existed even before money. People might have gotten sick and tired of rip-offs by fakers, and came up with the idea of money to try to regulate the problem. Fair-trading *without* money can be tricky enough even between two honest farmers. Having to establish the values of various goods all the time is a hassle.

Here is a theoretical example: "I'll give you fifty bags of potatoes for a cow," says a farmer to his neighbor. A faker farmer would probably respond: "Are you crazy? I had to raise this cow for three years cleaning up after her feeding her pampering her. You have to give me at least a hundred bags of potatoes." In reality, the potato farmer had to work his farm hard, while the cow was free to roam all her life causing hardly any trouble. The faker would blow things out of proportion, as he feels entitled to more. Naturally, he would try to squeeze as much out of the potato farmer as possible. Even then, he would still feel he didn't get enough. He feels he should get all the potatoes of the farmer for a cow. "What's that farmer doing with all those potatoes anyway?" – He thinks to himself. "He has more than what he needs". The faker has no regards for the labor of the potato farmer. As he would see it through his distorted self-elevated view, the "unfair" farmer wouldn't be likely to part from more than a hundred bags. (Unfair is a relative term here as the faker perceives it) Therefore, he tries to get the hundred bags not because he thinks that is fair but because his mind computes, that is how much he can get out of the farmer. He may even compute that he doesn't want to upset the farmer too much as he might be needing more potatoes next year. (Such computations don't take more than a split second for our mind, it only looks long when all written down). An honest farmer on the other hand would have said "C'mon now neighbor we've known each other for a long time. I know you worked hard on that farm, and this cow was hardly any trouble. Fifty bags are too much. Give me twenty five bags and we'll call it even." That's because the honest farmer considers the work of the potato farmer. (Others *do* matter) His mind makes a similarly quick computation but with a different result. His mind factors in the work of the other fellow and compares it to his own investment into raising the cow. Our mind does this so fast we don't even realize it. All that would register to the honest farmer is that fifty bags just doesn't feel right. Then his mind computes in another split second what would be right and comes up with twenty-five bags. Chances are that his computations will be spot on too. He noticed his neighbor worked hard. Our mind is in fact so accurate that in absence of false-information we could compute such equation down to fractional accuracy such as "twenty-four whole bags plus twenty two and a half potatoes". It is important to know this because even when we don't say anything about a deal gone sour, our mind computes what would've been fair. If that's not what we get we'll feel cheated. Chances are fakers have been messing things up for the rest of the population way before the monetary system. On a large scale, their activity creates an imbalance in the distribution of goods, causing scarcity and even poverty. These conditions will cause people to be suppressed and become needy, which in turn can make many of them struggle for survival. In such condition crime rises. All because of a simple rule deeply buried in their mind – rule#1.

For whatever reason, eventually the system of monetary exchange was accepted – this is a fact. In present time, there are organizations and individuals giving out loans and expecting outrageous returns on their investments. Fakers have been infiltrating and corrupting the monetary system for a long time. It's been going on for so long that recent generations rarely think about the unbelievable interests they pay to *legalized* loan sharks. Fakers achieved to make this an accepted way of life. Regulators are of course attempting to patch things up, but these will be temporary solutions until we cure the root cause.

One crucial aspect of money wasn't taken into consideration right from the beginning, or has been forgotten. (Or done away with by fakers) Money represents goods or services. Its value is based on, the amount of goods or services for which it can be exchanged. Money, however, are *not* goods. It's merely a *representation* of the assigned *value* of goods. Loans are a fundamental violation of the principals of money. Not that it matters much, because this is just a fundamental argument about the rules concerning money. Just passing more laws about it alone won't solve the problem. Fakers will find a way to trick people, no matter what the system, laws, and regulations are. This is what the quote from Plato tells us also, which was stated more than two millennia ago. We are generally unsuspecting of how our trust is being violated, thus unwittingly giving fakers the upper hand with or without the monetary system. The monetary system is a useful one, and simply changing the system won't solve the problem. For example in a fair world people wouldn't give predatory loans to begin with, even if the law allowed it. In our world, fakers will find ways to give predatory loans no matter how many laws we pass as long as we give them the power to do so.

Many systems have been tried, in our world, even socialism. Capitalism, socialism, whatever-ism can't make this problem go away. History is there to show this as well as current events in the world. Fakers are slippery and clever and therefore there is no possible way to filter them out by changing or improving the social systems or laws. Understanding and spotting them is the key to controlling them.

Understanding fakers could also be thought of, as an extension of the Ancient Greek aphorism: "Know thyself." By extending that to humanity as a whole, we get "Know ourselves". Which would mean that ideally every one of us should know what we (humans) can do, or become.

Fakers are part of us (humanity) therefore knowing about them, is to know our nature.

We live in a shared universe, so it is inevitable that people will often bump heads over material goods. This isn't so obvious in our everyday life. We have a currency system, laws, and law enforcement to protect and uphold property rights. There are beautiful homes, cars, artwork, and gadgets that many people would like to have, but only one person can have any one of these, at any given time. Once exchanged, it becomes someone else's property. We moderate the "head bumps" over material goods with money, laws, and regulations, and the enforcement of these.

We all differ in our needs, dreams, and wishes. Some prefer city life with its busy and fast-paced work style, while others prefer a more quiet and peaceful setting and less stressful work schedules. People aren't the same and thus even in a fair and honest society some will always end up with more property than others. If they work more, they deserve more. That is how it would be fair and there's no changing that. People aren't all the same, not at all! Some work more than others. It's way within the definition of fairness that those who worked more, should have more. The difference is that in a society without fakers, people who have less would not be upset about those who have more; because they would know, they had to work for it and achieved it fairly. Unfairness instilled by fakers is what rigs our game of life and causes upsets.

Fakers want more because they have a feeling of entitlement due to their twisted vision rooted in rule #1. Therefore, they *must* get more by whatever means, which creates a chaotic and unfair distribution of wealth. In the chaos they create, many of us get less than we deserve, which is a continuous source of upsets over the distribution of wealth throughout society. This upset would not be present in a fair system where fakers can't roam freely. Reaping the profits generated by others is not fair, and fakers do that very frequently, thereby perpetuating this material chaos.

Our systems for sharing are not bad at all. We know that it takes money to buy or rent things for use, and generally respect the property of others as much as we want others to respect ours.

Of course, because of rule #1, fakers can end up chasing money and power endlessly. This irrational greed for more, way, way more than what they need and deserve, never stops. The greed doesn't stop even when they have so much wealth, that they could live comfortably without lifting a finger, for a thousand years.

Their *false-known* rule #1 governs all of their actions concerning others, and as they spread chaos, they can instill irrational behavior in others as well. Therefore, they constantly throw off the balance of sharing. Fair sharing of material goods, involves the consideration of the wellbeing of others. Since fakers have irrational thoughts concerning others, their conduct in sharing will also be irrational.

This is the whole of the problem of sharing. If everyone was honest, then sharing would be no problem whatsoever, since greed is born from the distorted views of fakers. When we have a natural view that "others *do* matter", it's easy to fix our problems because there is a general consideration for the wellbeing of others.

Dealing with Obstacles

For any obstacle, there are only three possible courses we can take:

- Overcome
- Choose another path
- Ignore/give up

Let's get the option of ignoring/giving up out of the way first. These are not recommended. Imagine for instance that someone wishes to start a business, but ignores the fact that eventually $50,000 will be required to start. It is evident that ignoring this is a sure way to a persistent problem. Giving up is not recommended as it can negatively affect our self-confidence.

We will look at "overcoming an obstacle" and "choosing another path" all together.

For example, a student who has a plan for a business requiring $50,000 may be stuck thinking that his only option is to earn that money in order to overcome the obstacle. Examples of choosing a different path would be to get a loan or find a partner. Each choice has different outcomes.

Earning the money takes time and effort, but the upside is that we can do as we please. Borrowing will most likely mean interest payments and deadlines. If a partner is involved, we have to share the decision-making and profits. The key is to remain flexible and open to alternative solutions. The obstacles we face are not always possible, or feasible to overcome. In each case, we can realize and accept the need to change direction. If the obstacle is too time consuming, we can consider other alternatives. When dealing with obstacles it is best to focus our persistence on the final goal. If we have fixed ideas about how we're going to reach the goal and are not flexible in changing directions, we may be reducing our chances of success. When we set out to achieve a goal, there will be challenges. Focusing on the goal is the most important part. How we get there should be optional and kept open. That way we can adapt to circumstances, and use any opportunity presented.

For example, it would be inflexible of a hiker to look at a distant mountain peak and say, "That's where I am going, and I am going straight no matter what". Meanwhile, he has no clue if there are quick sands, swamps, or even lava on that path. When we hit such obstacles in life, it may be easier to go around them.

This is why it's important to set a goal and keep it as the highest priority. (Keeping an eye on the mountain) Everything else leading to it should be flexible, as we find the workable way. This allows us to find our personal walk in life.

This doesn't mean we should avoid every obstacle. Changing direction is for those obstacles that are nearly impossible or unreasonable to overcome. The idea is to select the ideal path. An obstacle that will require forty years, just to achieve one small step towards our goal should be examined for alternatives. Confronting obstacles and overcoming them is necessary. It is ok to hike a few hills and swim across some rivers, and climb a few trees on the path to the mountain.

Conclusion

History speaks for itself, so we don't need to go great lengths to prove that all the different methods of social control appear to fail sooner or later. Looking at this alone suggests that the real source of our problems is elsewhere. Most politicians, legislators, governors, lawmakers, law enforcers, and senators are trying to solve our problems, right here right now. Most of them are honest people who *do* care about us, and the future of our society. While others not so much. The honest ones are facing great problems and challenges, and are trying their best to fix them. Many of their efforts are in vain, as fakers work against them, day in and day out, waiting to *cash-out*. Fakers don't care how much corruption, chaos, and damage they cause.

Historically, people are willing to try just about anything, in hopes of a safe, sane, and fair society. Honest and hardworking people just want one thing from any system: fairness.

To make our society fair we should improve our ability to spot unfairness.

As we consider human nature, we can see that some people compulsively lie to get ahead. Such people are just as much part of humanity as any of us. They have the right to freedom and happiness but they need our help to guide them out of chaos. They can't help what they are doing, which is why it is necessary for us to step in. This book points them out, and presents theories of the underlying reasons that explain the source of their mentality. We have new and accurate tools to detect who should be trusted with power and who shouldn't. As we increase our collective awareness on these subjects, we'll also increase fairness throughout society. As fairness increases, our social problems will reduce.

Practical Application

The next chapters describe the practices we can use to find false-information, fakers and restore trust in our split second judgments. Practicing five to ten minutes daily is all that's required. By doing so, we are essentially telling our mind that these subjects are important to us. The daily presence of these practices in our lives combined with interest is all that's required in order for our minds to start adapting these simple techniques. The purpose of these is only to serve as a daily reminder until our mind adapts to the "idea". Then we'll be able to do these any time as required. The practices should be tailored to our specific needs and problems. The guides and rules about customizing the questions in the chapter *Persistent Problems* applies to all of the practices.

Finding False Information

An important aspect of these practices is to remain objective about past events. When false-information is present, we are not looking for "why" it is present. We already know that. It's there because somehow we got sold on it being true. We accepted it for some reason and that's that. Instead, we are looking for events and experiences, which contain false-information we have accepted. We fish for these mostly with words such as "who", "what" and "when". The purpose is to find false-information. (Remember feeling out of place or bogged down indicates the presence of false-information) The false-information is not readily visible to us. It is buried under the illusion or assumption, which caused us to accept it. Such as, "the assumption that the source was telling the truth", or "the source is always right" or simply because of the appearance of things. Therefore, we should look in the past as opposed to the now, and recall real events that took place. Something that was told to us, or done to us. Something we have experienced in real life. We are not fishing for imaginary or subjective thoughts. We are looking for "who" or "what" made or contributed to us having false-information.

We are very strictly fishing for false-information and false-information alone.

Any form of false-information as described before. Influence and manipulation is also based on false-information. Therefore, the influence of a faker will lose its grip on us when we spot the false-information he presented. We are looking for some fact that we have accepted that is incorrect. This is a fact that does not match up to reality. Even when we find one the recommended question isn't "why is that information false?" but instead "what makes that information false?" or "who made that information false?" For the duration and purpose of these practices, it is best to erase the word "why" from our vocabulary. (Not permanently just for the practice) The rules in the chapter *Persistent Problems* apply to all of the practices.

The best way to practice finding false-information is by making a list of facts connected to a problem. We can pick a recent or old persistent problem. Any problem will do. If we don't have one we can make one by looking up any field of study we don't know yet and get ourselves into some trouble intentionally. If that doesn't do it then we can ask someone else (friend, family, associates, colleagues etc.) if they have a problem we can help figure out. If that all fails we can even do fun stuff like addressing the issue, that we can't find a persistent problem to work on. To help make the list we can ask the following questions a couple times:

"What fact did I forget to list?"
"What fact could I add to the list?"
"What fact shouldn't I add to the list?"
"What fact must I add to the list?"
"What fact I must not add to the list?"
"What fact am I trying to hide from this list?"
"What fact is hidden from this list?"
"What fact has been skipped from the list?"
"What fact was I supposed to hide?"
"What fact isn't connected to the problem?"
"What fact is connected to the problem?"
"What fact can't be connected to the problem?"
"What fact is absolutely not connected to the problem?"
"What connected field did I not look at?"
"Think of another connected fact."
"Try not to think of another connected fact."

The purpose of such questions is to help our attention to break free from any one particular field we may be too focused on. Once you get the idea, you can make more questions. The idea is to make ourselves think of things we haven't thought of before.

Spotting Fakers

Once we have a list of facts we can start working on verifying them. You should add to the list at any time whenever you find a new connected fact you didn't see before. The verification doesn't have to be all at once. In many cases, it works out best to keep the list around for a while and work on the verification over time with patience. The verification should be done as objectively as possible. If there is an observation you can make for yourself, then never take a substitute for that. If you can't verify something by direct observation then get at least two sources which prove the fact beyond a doubt. To help raise curiosity the following questions can be used:

"Is this fact true and verified?"
"Who/what would make this fact false?"
"Who/what would make this fact true?"
"Where did I get this fact from?"
"Was an error made about the fact?"
"Is there an error in this fact?"
"Is something omitted about the fact?
"Is something missing about the fact?"
"Is something assumed about the fact?"
"Is something corrupted about the fact?"
"Is something falsified about the fact?"
"Is something hidden about the fact?"
"Is there a false-information connected to (…)?"
"Is there missing information about (…)?"
"Is there omitted information about (…)?"
"Is there an influence connected to (…)?"
"Is there a hidden influence connected to (…)?"

You can also question about the source of the fact. To stay objective look for who or what presented the fact to you.

"Is the source of the fact incorrect?"
"What would make the source incorrect?"
"Was the source motivated to present false-information?"

"Did the source present false-information intentionally?"

Note: Only one of the words should be used in any sentence at a time when they are separated by slash '/'. For instance, the question "Who/what would make this fact false?"

Should be first asked as:

"Who would make this fact false?"

Then as:

"What would make this fact false?"

You can play with putting the sentences into past tense also. For example:

"Is something hidden about the fact?"

Can be changed to:

"Was something hidden about the fact?"

Or:

"Has something been hidden about the fact?"

Or even:

"Did someone hide something about the fact?"

You can also replace the word "fact" with the item in question. For example:

"Is something hidden about the operating hours?"

The main rule is to keep it objective (factual) and at the "facts" layer, by avoiding "why" type questions.

(…) should be filled in with the subject in question.

For example:

"Is there an influence connected to (…)?"

Could be filled in with as follows:

"Is there an influence connected to my carrier choice?"

Spotting Fakers

Finding false-information may require some practice before it turns up any results. Then again, you might have a good catch the first day. These practices should be done like fishing, with ample patience, and for the love of the sport.

Example of Finding False Information

This chapter should be used in combination with the previous one as it demonstrates additional example questions and methods that can be used.

In the following example, we'll look at the a dilemma of the manager of a deli chain store. This chain has a central office. The general manager in this office notices one day that the sales of all ten stores have dropped 20%. He knows that prior to that they had some major changes in their advertising campaign. Naturally, he assumes that this last change must have something to do with the drop in sales. He sets out to research quickly and issues a reversal order to all the advertisement changes. The next month the sales are still down 20%. Then he starts worrying because it doesn't make sense. He talks to all the local store managers and figures that the new advertisement probably upset and turned away 20% of their customers. He is determined to find out why. As his research and questions do not reveal the reasons, he gets frustrated. He has a persistent problem. Whatever he does, the sales are not climbing back to where they once were. He has some partial success and increases sales by 2%. There is still 18% missing from where the sales we a few months back.

Then one day, he has to take a detour on the way home, and drives by one of the local stores at 7pm and notices something curious. Only the drive through is open and there is an unusually long line. But why – he wonders? Then he finds out that while he was on vacation his substitute manager passed a request for change of operating hours without much (or any) thought. The drive through hours didn't change but the dining area closes at 6pm now instead of 8pm. Due to the long line at the drive through, many of the returning customers just go somewhere else.

After a great sigh of relief, he makes the necessary changes to reinstate the original operating hours. Within two months, the sales go back to the normal range.

What happened here is that when he made calls to the local stores they would always answer the phone and say they are open. From their own viewpoints they were. They were there working and serving people through the drive through. There was no one to serve the returning customers who liked to sit down between 6-8pm to eat there. Of course, the sales were down.

While this was a trivial made up example, it shows that suspecting a particular source can blind us from seeing the real source. The manager was focused on the advertisement changes because that's what he thought was the cause of the problem.

Fortunately, there are ways to work around such tendency. The first thing to do when a persistent obstacle or problem presents itself is to find all the facts connected to the problem and make a list. A full and complete list, with anything and everything, that has to do with the problem. If it is a complex issue then it is a good idea to list on a large paper of drawing board. The next thing to do is make a list of all the things that we think are *not the source* of the problem. Then make another list with things we are *absolutely sure* not causing it. This is best done by asking ourselves simple questions such as:

"What part of the company is not causing this problem?" "Which part of the company has absolutely nothing to do with this problem?" "What changed that I don't know about?" "Which employee is the least likely to have caused the problem?" "What recent change is not related the problem?" "What policy or rule is not related to this problem?" (We can add the words "absolutely sure" to these questions)

Such questions can be asked at a company meetings also. This is not an exhaustive list of questions. The purpose of these is to illustrate the concept that we should "fish" for something we haven't looked at yet.

The source (cause) of a persistent problem is hidden.

How do we find something that is hidden from our sight?

By looking where we think, "it isn't".

The nature of persistent problems is that they persist because we have not examined the correct source. When we have an idea about where the problem is, our focus will be on the idea, which can make us blind to other possible sources. If the problem is not getting resolved (it persists) it is not because of some complex issue that requires fine tweaking. It is because we missed something obvious. As in the example, where the dining room was not open to take the money from the returning customers in exchange for food. Of course, the sales were down.

The purpose of making the long lists is that if the problem is persistent it means we have missing or false-information somewhere. Hence, we assume. Using the list helps to systematically go through and verify everything. The verification is crucial. If we leave room for deviation, we might not find the problem even with the list.

In the example, the manager could have made a long list of all the fact and logic connected to the stores. Starting with food delivery, when and how? Food being prepared, when, and how? Store opening and serving customers, when, and how? Operating hours, cleaning times, conditions of the stores, service, quality, food quality, store appearance, staff attitude, etc. The list can be long but it is a good idea for any business owner to make a list of all the factors that the business depends on anyways and keep these handy. Like a pilot keeps his checklist handy. That way we can do routine inspections on these. This example shows well that when making a list nothing should be grouped together. For instance, the operating hours of the drive through and the dining room should be handled as two separate items on the list. Thereby each requiring verification. A call to the store and asking if they are open is too general of a question.

These are much more specific questions:

"Is the drive through open?"
"Are there customers in the drive through?"
"Is the dining room open?"
"Are there customers in the dining room?"
"How many?"

Ten minutes before closing, such questions would have revealed the problem at ease. As trivial and silly as it sounds to check such basics, the source of persistent problems is usually trivial and overlooked issues.

We should check the listed items one by one to see if it could be broken down into elemental components. Each item should be looked at as a fact and we should list all of the dependencies of the fact. We can do these with simple questions such as:

"Does this (item) have a dependency?"
"Have I missed a dependency?"
"Is there a hidden dependency?"
"Is there a dependency I haven't thought of?"
"Can (item) be split into components?"
"Does it have a separate component?"
"Is there something (item) can be separated from?"
"Have I separated all?"

Again, these are merely sample questions. You can make up your questions as needed. The idea is to make up questions, which direct our attention to dissect as many elements of the problem as possible. The longer the list the better. Finding all the hidden and unspoken dependencies of facts is also crucial. We can't verify something that is hidden from our sight. If we focus on one area only then all other areas can hide from us.

It is equally important to mention that verification should be done to certainty. In the example, a personal visit to the stores would be the way to verify things in person and know beyond a doubt. This is going to be another trivial example: What if the sales were down 20% because of a construction project, which prevented the customers from entering the dining room after 6pm.

Spotting Fakers

A simple visit to the stores at various times would certainly reveal that the customers can't get in. Such trivial problem sources are the easiest to overlook. Example questions, which would uncover such issue, would be:

"Can customers get in at all operating hours?"

If you run into a complex and persistent problem, it can really help making such lists, and question them thoroughly. We can do this remotely at first using phone, Internet etc. If the problem still persists then it is time to get out there in person and investigate.

This process helps avoid confusions. We tend to get overwhelmed only when we take too big of a chunk of the problem all at once. Therefore overwhelm is never the sign of us being unable to solve the problem. It just means we are chewing off too much all at once, or that we have false-information.

When we set goals, we will encounter problems. Some of these may become persistent problems also. We can achieve goals by breaking each problem down to its most elemental components. Then we can focus on handling what we can do in one day's worth of work. A goal or problem, which requires years of dedication, can overwhelm anyone, if we look at the problems and obstacles ahead all at once. The reason is that anything that takes years to achieve is greater than what we can perceive. Just looking at it is overwhelming.

Problems are either simple or complex. We can dissect complex problems and break them up into smaller ones. Then we can address each step individually. This is why it's important to keep an eye on the goal, and not the long road ahead. We can perceive the final goal without being overwhelmed. This can give us inspiration daily. When we look at the goal, we should avoid pondering on the time and effort it will take to achieve. We should only concern ourselves with one day's worth at a time. After breaking down complex problems, there may be times that a simple but persistent problem arises. It may seem impossible to solve despite its simplicity. It's like a mountain on the road ahead that can't be climbed. It is natural to think that such problem requires more thought and effort. Think more; work harder. This is exactly what we <u>shouldn't</u> do.

The following may seem like an unusual approach at first. When a persistent problem arises, and it doesn't get resolved no matter how much we work at it, then the thing to do is STOP THINKING. Then turn "back" and start making lists of facts so that we can verify them and catch the false-information.

When we find ourselves voicing our inner thoughts with emphasis, it is a sure sign of a persistent problem. As described before, that's a sign of no trust over our mind, which is rooted in false-information.

The trick is to turn away from the problem (exact 180° turn) immediately, and start looking for false-information having to do with that problem. Examine every fact our knowledge is based on, one by one. For each one of these, we can ask questions like:

"Is the fact true and verified?"
"Did I skip verification of the fact?"
"What would make me skip verification of the fact?"
"Where did I get the fact from?"
"Who gave me the fact?"
"Who/What made me accept the fact?"
"Who gave me the idea of the fact?"
"What did I miss about the fact?"
"What is hidden about the fact?"
"What would make the source of the fact incorrect?"
"What would make the fact incorrect?"
"Who would make the source of the fact incorrect?"
"Who would make the fact incorrect?"
"Was the source of the fact false?"
"Was an error made about the fact?"
"What could make the fact false?"
"Was a lie accepted about the fact?"

These are just a few examples of questions we can ask to find false-information that causes the fact to give us trouble. We can make up others on our own. The key is to focus on the what, who and when and not use the words "why" and "how" in these questions. (See chapter *Persistent Problems*) It is important to find out "when" and from "whom" we received the fact. From there, we can see how it entered our reality. Who or what made us accept it. How does it deviate from reality is the key. Once we have that, the false-information becomes true because by spotting the deviation from reality our mind automatically corrects the fact to match reality.

We can rephrase the questions above by replacing the word "fact" with anything else we are examining. For example:

"Was a lie accepted about the fact?"

Can be changed to:

"Was a lie accepted about the money transfer?"

Or:

"Was a lie accepted about the operating hours?"

Another example:

"What is hidden about the operating hours?"

Of course, the question has to make sense to you and it should relate to the problem at hand.

It must remain objective.

If we don't keep it objective then we are likely to dig up imaginary stuff from our mind. We are not after our imagination here. We are after facts that do not match up to reality.

Additionally, we shouldn't make the question too general. For example, if you just simply ask:

"Was a lie accepted?"

Then your mind might dig up and give you some lie from childhood. That's not what you are looking for. You are looking to solve a present time problem so the question must contain the subject related to the problem. Such as:

"Was a lie accepted about the meeting?"

We should be relentless in our search, while drilling down as deep as necessary but not all in one day. Digging around our past memories and making long involved lists might be exhausting. If you feel like you need to take a break, you should. Come back to the problem the next day, or after a lunch-break. Whatever gets your mind off it for a while. When we face a persistent problem, a useful attitude towards facts is a systematic verification without assumption.

If a complex fact is encountered it should be taken apart into its elemental facts. Each elemental fact requires objective verification. Just like a pilot would do on his checklist. Look and see for *yourself* that every item checks out. When doing this we should be curious about each fact like a detective, because the culprit for our troubles is hiding right in front of us, "pretending" to be a useful and correct fact.

We can think about problems all we want, but more thinking will never result in a solution. The mind's prediction logic doesn't require extra time if all the necessary information is present an true. When the facts are clear and the false-information is all cleared up, our mind spits out the result in less than a second. No matter how long or complex the problem is. Important to note again that false-information covers missing and omitted information also. If we run into any of that then the answer is to go after it and research. Find out what it is.

When we think, and think, and think... then think some more until we finally solve a problem, and finally get it, we tend not to realize that what really happened is that we found the false-information.

It isn't important if the false-information was due to a mistake, omission, or intentionally falsified. When we solve a problem, it is *not* due to our logic somehow straightening up. Our logic is just fine, all the time. Finding false-information is what solves problems. Of course, we can say that this is still part of thinking. We would be correct to say that because our mind does that also. Finding false-information is part of our "thinking" process.

However, a differentiation is required here. When we are stuck on a persistent problem, our mind goes around in circles inspecting the same known elements we think are the cause. (Loop thinking) When we do that, we can feel overwhelmed. Listing the facts out on paper can help break the loop and help us extend our view wide enough to look at the real source of the problem. Once we extend our view then we can begin inspecting the facts one by one with great curiosity towards each.

Another simple but powerful question is, "Have I made an assumption about the source of the problem?" This should be self-explanatory.

Practices to Spot Fakers

These practices can help get started with spotting fakers. The practices are simple but strictly objective questions. The purpose is to spot fakers and find the false-information they presented. If we can see through the lies of fakers and spot all the false-information they spread, we become immune to their illusions.

These are the base questions related to sign 1:

Have I received out of place flattery today?
Have I observed someone else get out of place flattery today?

If you get a yes to any of the above questions, then it's important to ask the follow-up questions:

Who gave the out of place flattery?
Who received the out of place flattery?
What was the subject of the flattery?
What made the flattery out of place?

These questions are important for the following reasons: They keep things factual, so our mind doesn't run off to imagination land. Our mind has amazing, and essential, abilities of imagination, but it is best when we remain in charge most (or preferably all) the time over our creative imagination. For error free knowledge, we should call on the mind's other amazing ability, which is infallible precision.

We should keep these matters objective and not try to fill in the answers with guesses. While attempting to answer these questions, we may realize that the reason was imaginary, in which case we should dismiss it and resume looking for events that really happened.

The other reason for the follow-up questions is that by answering them we deepen our understanding of the situation, and build our practice and accuracy of recalling. Our subjective reality will improve accordingly.

Example questions and answers:

Who gave the out of place flattery?
Jane.

Who received the out of place flattery given?
Joe.

What was the subject of the flattery?

Jane complemented Joe on his personality more than once during a short conversation.

What made the flattery out of place?

Jane did not know anything about Joe because they just met for the first time. Therefore, she could not honestly give these compliments. They were just talking business. Every time she gave him a compliment, I felt weird as if I didn't know where to put it.

It may take some time and practice at first to think these through, so patience is important. In the beginning, the whole subject may appear involved but the more we practice and think about it the simpler it gets just like it is with any other involved skill. Every outpoint we spot makes us more aware, even if we don't spot a faker. The ability to spot outpoints is *priceless*. Just think about great standup comedians. They can spot outpoints in life then get up on stage and put it in a form that makes our ribs hurt. (From laughter) In the beginning, even if we get a total blank (no answer) on the follow-up questions it's no problem at all. Overdoing it, or becoming inpatient is never helpful. Asking "why" we didn't have an answer is not a good idea. We know why. It's a new subject and we need practice. If we didn't get a response it means we didn't have one available. It doesn't mean anything else. These practices are meant to be just done as is. When we don't have an exact answer it's best to just skip it. Chances are an answer will pop up in our mind later but even if it doesn't then the next time we do the practice it will go a little easier. Repeat the follow-up questions a maximum of three times. If you're still getting blanks, just go for the next subject or end off for the day. Getting back to it routinely on a daily basis is what's best, as this is what communicates to our mind that this is, in fact, a required skill now. Hitting blanks should not be seen as discouraging. It's perfectly normal to have a hard time at first if we have no experience and practice answering such questions. Getting blank at first is as normal as hitting our fingers with a hammer a couple times, when we first learn to drive a nail into a two by four.

An important detail is we can change the duration on the base question as we see fit. Say you had a busy week, and didn't practice for a week. No problem. Just change the word "today" to "this week", "last week", "this month" or whatever works. You can even say "Since my last practice", or you can simply omit the word "today" entirely, which extends the question to any time. Don't worry, your mind knows language, and it knows the difference automatically. What may be surprising is how willingly and accurately we can answer these questions. The key is not to dwell if we don't have an answer, instead just go for the next question. No answer just means we have no issue related to the question for the time being.

Examples:

Have I received out of place flattery this week?
Have I received out of place flattery since my last practice?
Have I observed someone else getting out of place flattery in the past two days?

These sentences are phrased to address a broad range of subjects. This is essential so that our mind can respond to anything that applies. If there is no response, don't push it or dig around too much. (Especially don't ponder "why") Just practice daily if you can, but keep it short and pleasant. This is supposed to be fun. This is not a self-interrogation practice. If you stray from the subjects, you are in uncharted territory, and on your own. The key is to keeping things factual. We should stay away from assumptions, guesses, and fantasyland. We should work with facts that we can know to be true, similar to the example given. (Jane and Joe) When your mind gives you a response, don't second guess it. Even if it turns out that the response is inaccurate, that only means that the response itself is also based on false-information.

Having to give answers doesn't mean to voice it in our mind. For example, a lengthy response to the *"What made the flattery out of place?"* can play out in our mind in a split second. We don't have to voice it piece by piece. When you know, you know! We want to encourage our mind's ability to know in a split second. Second guessing it, and forcing it to voice is not recommended.

Verification questions:

The purpose of the verification question is to see if we can find earlier times, of the same event, which can help discover an already existing pattern if there is one.

In the case of the previous example, the verification question would be:

Did I observe Jane giving out of place flattery before?

Or

Did I observe Jane giving "out of place flattery" to anyone else?

The idea here is to stay with the subject, who we have just observed to possess a sign of fakers, and see if we can find similar events. This can help us in more than one way. First, it allows us to practice digging up events from our memory and prove to ourselves that we can actually do that. (We all can) Secondly, we can observe that our mind is very willing to answer questions like that. The last and equally important aspect is that we can determine if there is a pattern, or if it's just an isolated incident. People may occasionally pick up and copy the habits of fakers, which result mostly in isolated faker-like behavior. We are looking for a repeated pattern with most of the faker signs. Only that means that we spotted one.

Spotting Fakers

For instance, in the example Jane might have just been simply attracted to Joe, which made her want to make an impression. Perhaps a habit she picked up by following the advice of a faker "friend". If we find that Jane only does this with a man, whom she later admits she was attracted to, would instantly dismiss her from being a faker suspect. Especially if we find she doesn't match any of the other signs. For instance, she is very cheerful and hardly ever has bad news. In fact, it seems to pain her if she has to share something negative. Never badmouths anyone either. This means that she is clean on Sign #5. Then you can dismiss her almost certainly. The signs should be observed in combination, and reoccurring.

Pondering on the results especially with "why" and "how" should be avoided at all costs. It is better to end off and come back to it later than allowing our mind to "loop think" with "why" and "how". When you catch your mind wondering, you can even give yourself (your mind) some commands like "Let's stay focused on actual stuff" or "Let's keep it real" or "Let's get back to the subject". You can and should do whatever works for you to keep you focused.

For every practice result, there is one more crucial step to do. Find the underlying false-information. Once we spot all the false information connected we regain our total freedom from that issue. That's how we can make sure that we are unaffected by the attempted illusion. Since there can be more than one false-information connected, this is a repetitive process until we realize that there is no more connected false information. We can use the same questions as in the *Finding False Information* chapter. In case of the Jane and Joe example after the "*What made the flattery out of place?*" question we could ask:

What false information is connected to the flattery?

An example response would be:

Jane had false-information of the missing kind about Joe because she does not know him. She gave Joe made up compliments. That is false-information presented as truth. In addition, Joe didn't seem to notice this. Joe received the lie and accepted it. Joe has missing information about Jane also. He does not know why she complimented him. Joe made the assumption it was due to attraction because of the smiles. He doesn't know. He assumed. Potential false assumption.

The idea is to look around and analyze all the involved parties and attention paths. (From chapter *Attention Paths*) Using logic and sticking to facts, we can figure a whole lot from such a seemingly trivial event. Once we spot all the false-information we'll know because we'll feel released. It is a sensation we get when we free ourselves from the load the illusion causes. Even if the illusion isn't meant for us, as long as we are present and observing it, it will affect us also, until we spot it.

This is the end goal of each practice. To feel relieved, cheerful and certain. This will register as feeling good about ourselves.

Once we have that, we should enjoy this state and take a break or continue the next day.

Sign 1: Out of place Flattery. Base questions:

Have I received out of place flattery today?
Have I observed someone else get out of place flattery today?

Sign 1: Out of place Flattery. Follow-up questions:

Who gave the out of place flattery?
Who received the out of place flattery?
What was the subject of the flattery?
What made the flattery out of place?

Sign 2: Know it All. Base questions:

Did anyone appear to know it all today?
Did anyone appear to know a subject better than anyone today?
Did anyone appear to have unquestionable authority today?
Did anyone get upstaged today?

Sign 2: Know it All. Follow-up questions:

Who was it?   (This gets the '(name)')
What was the subject, in which '(name)' appeared to know it all?
What was it that made me suspect, that '(name)' may not know as much as he/she claimed?
What do I need to do in order to verify if the claims of knowledge are valid or not?

Remember, you are looking for someone who claims to know more than seems possible. The *'(name)'* is replaced with the name of the person in question. Also note that the word "how" is not used for a good reason even if it would make for a simpler sentence. Instead, "What do I need to do" is used in order to stay objective.

Sign 3: Bearer of Bad News. Base questions:

Did someone give negative remarks to me about someone else today?
Did someone get badmouthed today?
Did someone's skill get questioned today?
Did someone's integrity get questioned today?
Did I see anyone tell negative remarks to someone else today?
Did anyone present bad news with enthusiasm today?

<u>Sign 3: Bearer of Bad News. Follow-up questions:</u>

Who was badmouthing whom?
Who was listening to the negative remarks?
Who was the negative remark about?
What was the subject of the negative remark?
Was it negative remark or a valid complaint?
What do I need to do in order to verify if it was a valid complaint?
Who was bringing bad news? (This gets the '(name)')
What was the bad news?
Did '(name)' ever bring good news before?
Does '(name)' seem to always bring bad news?

Obviously you only ask the follow-up questions which apply. For example, if you found someone badmouthing on a base question, then you only use the follow-up question(s) related to badmouthing. Remember you can (and should) tailor these questions or make up new ones as long as you keep the rules and stay strictly objective. The idea is to develop your own ability to come up with the necessary questions that fit the occasion.

<u>Sign 4: Not Answering the Question. Base questions:</u>

Did anyone have a suspiciously long delay after a question today?
Did anyone avoid a question today?
Did anyone ignore a question today?
Did anyone misdirect today instead of responding?
Did someone give no response to a question today?
Did someone repeat a question, several times today, to get an answer?
Did anyone pretend to be upset to avoid a question today?
Did anyone just simply leave to avoid a question today?

<u>Sign 4: Not Answering the Question. Follow-up questions:</u>

What was the question?
Who asked the question?
Who was the question directed to? (This gets the '(name)')
What did '(name)' do to avoid the question?
Was there a rational reason to avoid the question?
What do I have to do to verify if it was a rational reaction to avoid the question, or not?

Note: In any of the questions where it makes sense, you can replace word "anyone" or "someone" with "I' or even "we" instead. These may appear essentially the same, but in reality, we may get different answers with each change. Also if we repeat a question unchanged we may get a different answer every time we ask it. Our mind is like a very efficient search engine in this sense. It can return the most relevant response. Once we get a response, it tends to give us the next most relevant response when we repeat the question.

<u>Sign 5: False Accusation. Base questions:</u>

Did I get falsely accused today?
Did someone get falsely accused today?
Did someone not trust me today?
Did someone else not get trusted today?
Did someone mistrust today?
Did someone show lack of trust today?

<u>Sign 5: False Accusation. Follow-up questions:</u>

Who was accused?
Who was accusing whom? (This gets the '(name)')
What was the subject of the accusation?
What made the accusation false?
Does '(name)' accuse others frequently?
Who was not trusted?
Who was not trusting? (This gets the '(name)')
What was the subject of trust?
Does '(name)' often mistrust others?

<u>Sign 6: Feeling Entitled. Base questions:</u>

Did I see anyone today, who always needs to be contributed to?
Did I see anyone today, who seems to never get enough?
Did I see anyone today, who thinks they deserve it all?
Did I see anyone today, who thought they deserve more than they worked for?

Note: You should keep a thesaurus handy when you practice forming your own questions. You may be surprised how much the phrasing can affect the results. For instance, instead of the word "see" you can use, observe, detect, sense, witness or any other similar word that you feel comfortable with. As long as the synonym is an objective word, you can use it. Avoid words like "feel" for instance. It is not nearly as objective as "detect" for instance. The word "feel" may steer you to the "known" layer. Anything that steers or keeps us on this layer should be avoided for the practices. When you make your own question, it is important to check each word you use individually and ask yourself if it is objective enough. If in doubt, don't use it. If it doesn't seem right then it probably isn't.

238

<u>Sign 6: Feeling Entitled. Follow-up questions:</u>

Who was it?   (This gets the '(name)')
What was the subject?
What did '(name)' say or do, which indicated they deserve more/all?
What did '(name)' say or do that indicated he/she needs to be contributed to?

<u>Sign 7: Reporter of Conflict. Base question:</u>

Was a personal conflict told today?
Did someone complain about a deal gone sour?
Was only one side of the story told today?
Did someone omit something about a story today?
Did someone tell an altered story today?

<u>Sign 7: Reporter of Conflict. Follow-up questions:</u>

Who was telling the conflict? (This gets the '(name)')
What was the conflict about?
Who was the other party in the conflict?
Was the other party made to look bad by '(name)'?
Is this the first time the story was told?
Was the story told as a preemptive attack?
What part of the story can be lies?
What part of the story can't be lies?
What do I need to do to find out the other side to the story?

Sign 8: Faker vs. Faker. Base questions:

Did anyone argue today, while rapidly changing subjects?
Did anyone try to upstage another today?
Did a know-it-all meet his match today?
Did two fakers have an argument today?
Was someone upstaged today?
Did someone get personal instead of using logic today?

Sign 8: Faker vs. Faker. Follow-up questions:

Who was arguing?
Were they coming off as authorities or know best on multiple subjects?
Did it seem unrealistic that they could know so many subjects?
Did they seem to avoid answers by changing subjects?
Did they use responses that were irrelevant to the subject?
Did they resort to personal insults or references?
What did they do to upstage each other?

Sign 9: Leverage Seekers. Base questions:

Did anyone gain leverage over me today?
Did anyone gain leverage over someone else today?
Did anyone attempt to gain leverage over me today?
Did anyone attempt to gain leverage over others today?
Did anyone make me look weak or unstable today, by altering facts or information?
Did anyone make someone else look weak or unstable today, by altering facts or information?
Did anyone pretend to be friendly in order to gain leverage today?

Sign 9: Leverage Seekers. Follow-up questions:

Who was trying to gain leverage?
Who was gaining leverage?
Who was the target?
What information or fact was used to gain leverage?
By what means did the person gain the knowledge of that information or fact?
Was the person seeking leverage using false emotions?
Was there fake friendliness?
What made the friendliness fake?

Sign 10: Knowledge Bullies. Base questions:

Did anyone try to prevent me from finding out some information or knowledge today?
Did anyone try to prevent someone else from finding out some information or knowledge today?
Did anyone use opinion to steer me away from any information or knowledge today?
Did anyone use opinion to steer someone else away from any information or knowledge today?
Did anyone use non-creative criticism to stop some information or knowledge from reaching others?

Hint: opinion can be changed to authority.

<u>Sign 10: Knowledge Bullies. Follow-up questions:</u>

Who was it?   (This gets the '(name)')
What information or knowledge was '(name)' trying to stop from me or others?
What opinion/authority was used?
Who or what was the target of the non-creative criticism?
What made the criticism non creative?
What would make someone want to prevent others from knowing about such information or have such knowledge?

<u>Sign 11: The Spiel. Base Questions:</u>

Did someone tell a long story today that I had doubts about?
Did I read a story today that I had doubts about?
Was the moment of doubt today related to a story?
Did someone try to sell me something with a spiel today?
Have I heard a spiel today?

<u>Sign 11: The Spiel. Follow-up questions:</u>

Who gave the spiel?
Which story did I doubt?
What was it that made me doubt?
When was it that I doubted the story?

There may be more than one time during a story, so this is a good example of a question that can be repeated.

What did the story teller do to make me doubt?
What did the story teller do to make me believe?

Spotting Fakers

The purpose of this list of questions isn't to give an all-inclusive list of questions. There is no such thing.

The purpose is to demonstrate the idea through examples.

Many more questions can be made in the same fashion but the idea is just to have enough materials to practice with, which in turn will let our mind know that these types of observations are going to be required in the future. Remember no "why" or "how' should be used in any of the follow-up questions either. These can lead to endless pondering instead of revealing facts. After ample practice, these questions won't be required. You will be able to make up your own without even voicing them throughout the day. You'll just know like you know how to read. Once you realize that you are spotting signs without even thinking about them throughout the day then your practice is complete. Or if you just feel like your practice is complete then it is. You are the final authority.

The questions should be asked silently in our mind but could be said out loud also if that feels more natural or effective. Whatever works as long as you stay objective.

The important part is to really ask ourselves with the intention to get an answer.

Being curious about the response can really help with that.

Intention is not the same as force. Intention in this case means to ask the questions in a way that deserves an answer. The questions are directed from ourselves, to ourselves. So we should ask ourselves these questions directly. The practice should be done in a quiet and comfortable space, with no distractions. When such environment is not available, it's better to just skip a day than trying to do the practice while being distracted.

Spotting a faker, the influence a faker had on us, false-information, false assumption, omitted information, false know or anything that we didn't realize before makes us feel good about ourselves. That feeling means that we just got one step closer to recovering the infallible nature of our mind. When that happens it's time to enjoy our victory and take a break or be done for the day.

Enjoying the victory is important. This is what it's all about: To feel better, and be more able.

This is probably obvious by now but it should be said anyway:

This book is *not* a promotion to label or identify others as fakers. When we suspect someone to be faking, what is recommended is to keep it to ourselves and use that knowledge to make more accurate judgements about our future course and decisions. Publicly labeling others as fakers will only cause conflict. This book is not written to cause more conflict. We already have enough of that in the world. Inaccurate labeling can cause conflicts and confusions. Of course, everyone has their freedom of choice to do as they please. Those who still label others as "fakers" out loud or in their mind may have missed the points of this book.

Labels and judgements are sensitive subjects. Instead of labeling people as fakers, it is better to just think of them as *someone who shows the signs of fakers*. People can change and circumstances (especially false-information) can cause us to make inaccurate judgments. For this reason it is recommended to keep an open mind and be willing to re-inspect our own judgements as well.

When we spot someone using the signs, instead of thinking of the person as "a faker" we can (and should) still think of the person as an individual we rather not trust. Also we can keep our mind open just by adding "until I find otherwise".

People are individuals and not labels. This book is in no way the promotion of identifying individuals by a new (or old) label/term. Quite the contrary! When this book is used as intended to spot false information, it can help us break away from labeling and seeing people for who and what they really are – Individuals.

Increasing Trust in Our Split Second Thoughts

When we do spot a faker, almost without fail we can realize with a little poking around in our *mental-copies* (memories) that we had a negative first impression, which we eventually dismissed. It's very useful to dig up that first impression, and acknowledge its validity. It will validate the amazing split second accuracy of our mind. This way we can steer towards using and trusting these thoughts. While we are at it we can also check (by asking ourselves about it) who or what and when made us override that split second thought. (Not all in one question though)

The question is simple:
"What was my first impression of '(name)'?"

'(name)' should be replaced with the name of the suspected faker.

There are different combinations of this question:

"Was there a first impression of '(name),' which I ignored?"
"Did '(name)' ever make me feel out of place?"
"Did '(name)' ever make me feel uncomfortable?"
"Did '(name)' ever make me feel like something is not adding up?"

Using the word "feel" is ok here because we are looking for the feeling. We can use the word "sensation" also. Once we spot the time of the "feeling" we need to switch back to objective mode again and look for the facts that caused the feeling.

We can make up different combinations of these questions, but they are accompanied by some important rules. Don't wander off the subject too much with the questions, otherwise you may get lost in a jungle of assumptions.

We are not looking for assumptions here. We are working to get rid of the ones we already have. If you spot a false assumption, by spotting what made the assumption false during your search, that's a different story. Such should not be confused with assumptions made during the practice. When you spot a past assumption, and find who or what made it false, be sure to acknowledge your success to yourself. Questions, during practice (or everyday life) which lead to guessing, are anything but useful. This is another reason why the words "why" and "'how" isn't useful during practice. They invite assumptions. It is important to keep the questions objective. The purpose of the questions is to help us find real people, real events, and factual reasons.

The next important step is to ask what that first impression was.

Just feeling bad is not useful by itself, because we can have different irrelevant phenomena at work. Perhaps the person reminded us of someone we don't like, or stirred up the memories of a nightmarish childhood teacher, or the person's voice sounded like our least favorite actor or the perfume the person was wearing reminded us of hospitals. The examples go on and on, how we can get totally unrelated feelings and impressions when we meet someone. It is important to stay focused and stick to the subject.

Therefore, we can only use the response to such question if it's relevant to the subject of faker signs, as well as to the suspected person. If the response is something unrelated then we can try repeating the question until we can see if it brings up something feasible. We can play with the question around a bit, but again, we don't want to get into self-interrogation. If we get thrown off the subject (our mind wonders off), then we should come back to the question as soon as possible. The more we insist and practice, the easier it gets to catch our mind from wondering off the subject. For this reason, you may want to keep the book opened at the question, to serve as a reminder. Our minds can do some wonders, jumping from one subject to another, when we start asking questions of ourselves. This isn't a useful process, and is the result of false-information being present. This is why we should have the question right in front of us to help us remain focused.

Such questions can also be asked about other subjects, in order to recover our trust in ignored split second thoughts. As described before, these split second thoughts are just as much subject to being false due to false-information as voiced thoughts.

Every ignored split-second thought contains the false-information that the thought was incorrect, because we ignored it. When we spot that and validate the thought as correct, we handle the false-information. With each split-second thought we recover we get a little more able, efficient, and certain. It's understandable that we'll feel good about ourselves, and we should. With each step, we get closer to realizing what beautiful minds we all have.

Please visit www.spottingfakers.com where more information and discussions will be available.

www.ingramcontent.com/pod-product-compliance
Lightning Source LLC
Chambersburg PA
CBHW052034090426
42739CB00010B/1910